Additional Skill and Exercise Manual

Melanie Fulton

High Point University

College Algebra in Context

with Applications for the Managerial, Life, and Social Sciences

Third Edition

Ronald J. Harshbarger

University of South Carolina – Beaufort

Lisa S. Yocco

Georgia Southern University

Addison-Wesley
is an imprint of

PEARSON

Reproduced by Pearson Addison-Wesley from electronic files supplied by the author.

Publishing as Pearson Addison-Wesley, 75 Arlington Street, Boston, MA 02116.

Printed in the United States of America.

ISBN-13: 978-0-321-56970-7
ISBN-10: 0-321-56970-9

1 2 3 4 5 6 BB 10 09 08

Addison-Wesley
is an imprint of

PEARSON

www.pearsonhighered.com

TABLE OF CONTENTS

Dedicated to my husband and children:
Sandy
Bethany
Will

Section 1.1 Functions and Models

1. The table below gives y as a function of x, with $y = f(x)$.

x	-2	1	4	7	10	13
$f(x)$	-8	1	10	5	3	9

a. Is 5 an input or an output for this function?
b. Is 4 an input or an output for this function?
c. Is $f(-2)$ an input or an output for this function?
d. State the domain and range of the function.
e. Explain why this relationship describes y as a function of x.

2. The table below gives y as a function of x, with $y = f(x)$.

x	-5	-1	0	6	9	15
$f(x)$	-2	4	1	7	2	-6

a. Is 0 an input or an output for this function?
b. Is 4 an input or an output for this function?
c. Is $f(6)$ an input or an output for this function?
d. State the domain and range of the function.
e. Explain why this relationship describes y as a function of x.

3. For each of the functions $y = f(x)$ described below, find $f(3)$.

a.

x	-1	1	3	5	7
$f(x)$	0	4	8	12	16

b. $f(x) = 2x^2 - x$

c. $y = 3x + 6$

4. For each of the functions $y = f(x)$ described below, find $f(-1)$.

a.

x	-1	0	2	6	9
$f(x)$	0	5	4	17	12

b. $f(x) = 4x - 2x^2$

c. $y = 5x + 3$

5. If $P(x) = 9 - 2x$, find $P(4)$ and $P(3)$.

6. If $C(x) = 3x^2 - 6x + 5$, find $C(20)$ and $C(-10)$.

7. If $f(x)=\dfrac{3x}{2x+1}$, then find

a. $f(5)$ b. $f(-2)$ c. $f\left(\dfrac{1}{6}\right)$

8. If $f(x)=\sqrt{x+2}$, then find

a. $f(7)$ b. $f(0)$ c. $f(-1)$

9. Determine if the graph below indicates that y is a function of x.

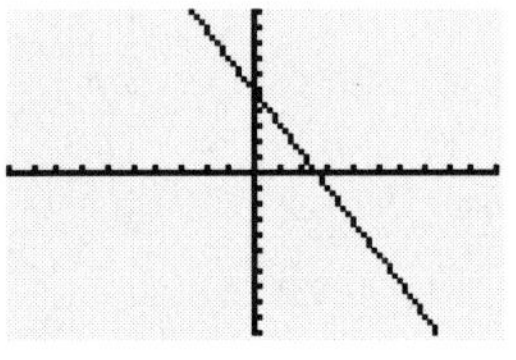

10. Determine if the graph below indicates that y is a function of x.

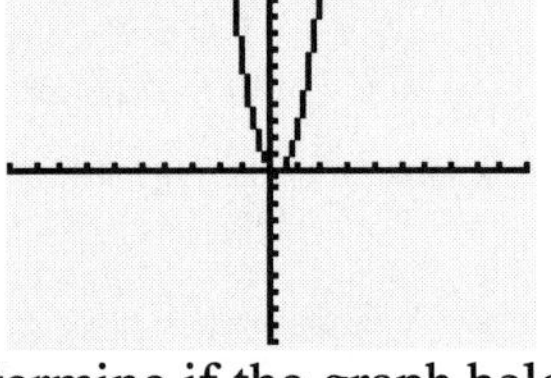

11. Determine if the graph below indicates that y is a function of x.

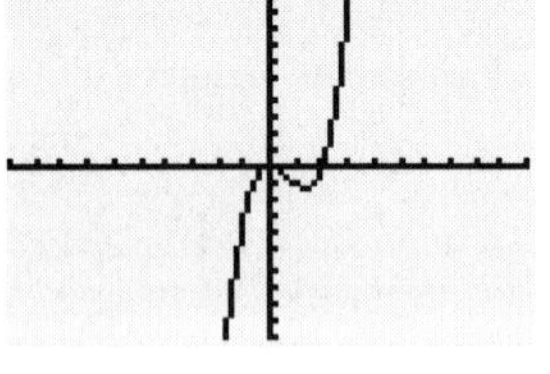

12. Determine if the graph below indicates that y is a function of x.

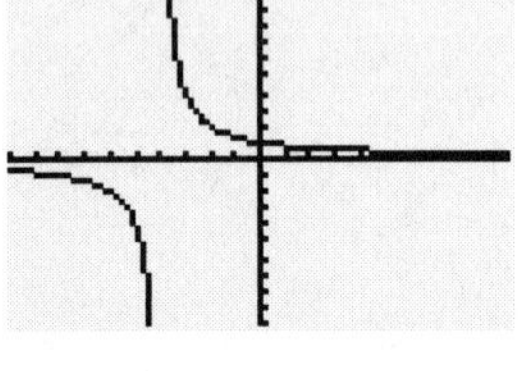

13. Determine if the scatter plot represents y as a function of x.

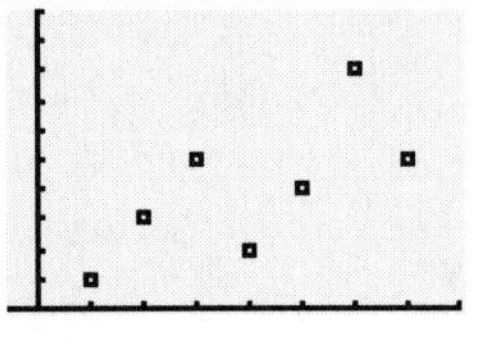

14. Determine if the scatter plot represents y as a function of x.

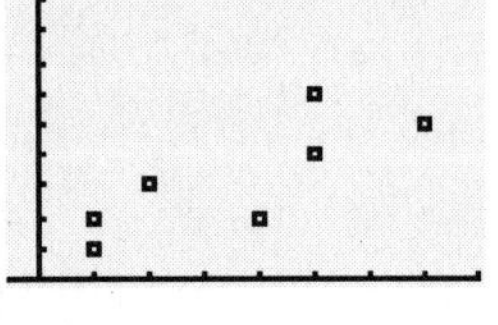

For Problems 15 – 18, determine if the set of ordered pairs defines a function.

15. (2, 3), (4, 6), (6, 9), (8, 6).

16. (1, 6), (-3, 5), (-3, 4), (-5, 3).

17. (2, 3), (4, 3), (6, 3), (8, 3).

18. (1, 4), (1, 6), (1, 3), (1, 9).

19. Does $y = 4$ describe *y* as a function of *x*?

20. Does $x = -6$ describe *y* as a function of *x*?

21. Does $x + y^2 = 1$ describe *y* as a function of *x*?

22. Does $x^2 + y = 1$ describe *y* as a function of *x*?

23. Does $x^2 + y^2 = 1$ describe *y* as a function of *x*?

24. Write an equation to represent the function described by: "The area of a triangle is equal to one-half the base *b* times the height *h*."

25. Write an equation to represent the function described by: "The average wage W is found by quadrupling the quotient of X and Y and then adding it to the difference of C from D."

26. Find the domain and range of $y = x^2 - 2$.

27. Find the domain and range of $y = \sqrt{4x}$.

28. Find the domain and range of $f(x) = \dfrac{2}{x+4}$.

29. Find the domain and range of each of the following functions.
    a. $f(x) = \sqrt{4x-8}$　　b. $f(x) = \sqrt[3]{x+6}$

30. Find the domain of each of the following functions.
    a. $f(x) = \dfrac{2x+1}{x-4}$　　b. $f(x) = \dfrac{2x}{(x+1)^2}$

Section 1.2 Graphs of Functions

For Problems 1 - 9, graph the following functions with a graphing utility using a standard window.

1. $y = x^2 - 4$
2. $y = 9 - x^2$
3. $y = x^3 - 3x^2 + 2$
4. $y = 2(x-1)^3 + 4$
5. $y = 3(x+1)^2 - 5$
6. $y = \sqrt{x-4}$
7. $y = \dfrac{3}{x-5}$
8. $y = \sqrt[3]{x}$
9. $y = \dfrac{1}{(x^2-4)} + 2$

10. Complete the table of values for the function $y = 3(x+1)^2 - 4$.

x	-3	-2	-1	0	1	2	3
$y = 3(x+1)^2 - 4$							
(x, y)							

11. Complete the table of values for the function $y = \dfrac{4}{x+1}$.

x	-3	-2	-1	0	1	2	3
$y = \dfrac{4}{x+1}$							
(x, y)							

For Problems 12 – 16, graph the given function with a graphing calculator using (a) the standard viewing window and (b) the specified window. Which window gives a better view of the function?

12. $y = 0.05x + 0.3$ with x min = -2, x max = 2, y min = 0.2, y max = 0.4.

13. $y = 20x + 50$ with x min = -3, x max = 3, y min = -10, y max = 150.

14. $y = 0.01x^2 - 0.02x - 0.15$ with x min = -10, x max = 10, y min = -0.16, y max = 1.05.

15. $y = \dfrac{0.2x(x-1)}{x^2 - 250}$ on [-10, 10] by [-0.12, 0.0196].

16. $y = -20x^3 + 160x$ on [-3, 3] by [-180, 180].

For Problems 17 – 21, find an appropriate viewing window for the function, given the x-values provided, and then graph the function.

17. $y = -x^2 + 25x$ for x-values between -5 and 5.

18. $y = \dfrac{-x(x-200)}{0.01}$ for x-values between -5 and 5.

19. $y = \dfrac{-x(x^2 - 64)}{0.1}$ for x-values between -10 and 10.

20. $y = (x + 200)^3$ for x-values between -205 and -195.

21. $y = (x - 1200)^2 + 1500$ for x-values between 1100 and 1300.

22. $y = (x + 20)^4$ for x-values between -5 and 5.

23. Find a complete graph of $y = x^3 - 4x^2 - 20x + 48$ showing two turning points.

24. Find a complete graph of $y = x^3 - 45x^2 - 1900x + 60{,}000$ showing two turning points.

25. Find a complete graph of $y = 2x^3 - 12x^2$ showing two turning points.

26. Use a calculator to find $S(t)$ for the values of t given in the following table.

t	$S(t) = 4.2t + 5.3$
15	
17	
19	
21	

27. Use a calculator to find $S(t)$ for the values of t given in the following table.

t	$S(t) = 14t^2 - 6t + 2$
20	
30	
40	
50	

28. Enter the data into lists and graph the scatter plot of the data, using the window [6, 35] by [10, 30].

x	8	12	16	20	24	28	32
y	12	23	18	20	14	17	25

29. Enter the data into lists and graph the scatter plot of the data, using the window [0, 40] by [100, 250].

x	5	10	15	20	25	30	35
y	120	140	160	180	200	220	240

For Problems 30 – 33, use the table below.

x	4	7	10	13	16	19
y	-2	5	11	19	26	36

30. Use a graphing utility to graph the points (x, y) from the table using a reasonable window.

31. Use a graphing utility to graph the equation $y = 2.49x - 12.75$ on the same set of axes as the data from the table using the viewing window [0, 20] by [-5, 40].

32. Do the data points fit the graph of the equation?

33. What is the connection between the x and y-values in the table?

For Problems 34 – 37, use the table below.

x	10	15	20	25	30	35
y	120	140	160	180	200	220

34. Use a graphing utility to graph the points (x, y) from the table using a reasonable window.

35. Use a graphing utility to graph the equation $y = 4x + 80$ on the same set of axes as the data from the table using the viewing window [0, 40] by [100, 260].

36. Do the data points fit the graph of the equation?

37. What is the connection between the x and y-values in the table?

38. Suppose $f(x) = 2x^2 + 10x$ thousand dollars are earned, where x is the number of years after 2000.
    a. What is $f(5)$?
    b. The answer to part (a) gives the number of thousand dollars earned for what year?

39. Suppose $f(x) = x^2 + 2x + 5$ fish are in a lake, where x is the number of years after 2000.
    a. What is $f(10)$?
    b. The answer in part (a) gives the number of fish in the lake for what year?

Section 1.3 Linear Functions

1. Which of the following functions are linear functions?

a. $y = 6x + 7$ b. $y = 2x - 3$ c. $y = 4 + \frac{3}{x}$

d. $y = 4x^2 - 5$ e. $\frac{3}{x} + \frac{4}{y} = 1$ f. $\frac{x}{4} + \frac{y}{2} = 2$

For Problems 2 - 10, find the slope of the line through the given points.

2. (5, 1) and (4, 9) 3. (-2, 4) and (-3, -1) 4. (5, 2) and (7, 2)

5. (-5, -9) and (0, 4) 6. (4, 3) and (4, -2) 7. (-7, -2) and (-5, -4)

8. (4, -5) and (23, 19) 9. (0, 150) and (10, 300) 10. (40, -20) and (50, -65)

11. Based on the graph of the line, determine whether the slope of the graph is positive, negative, 0, or undefined.

a.

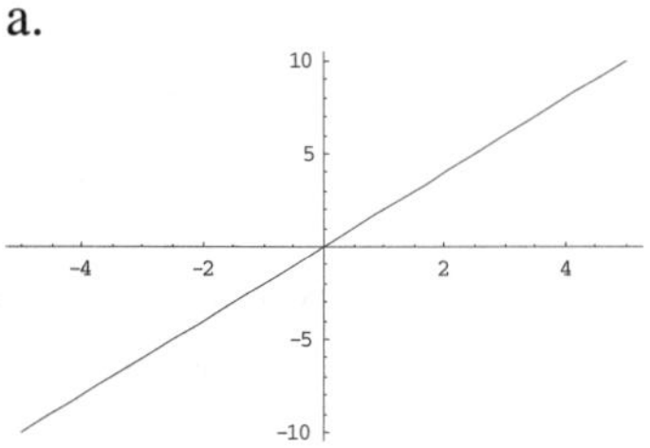

b.

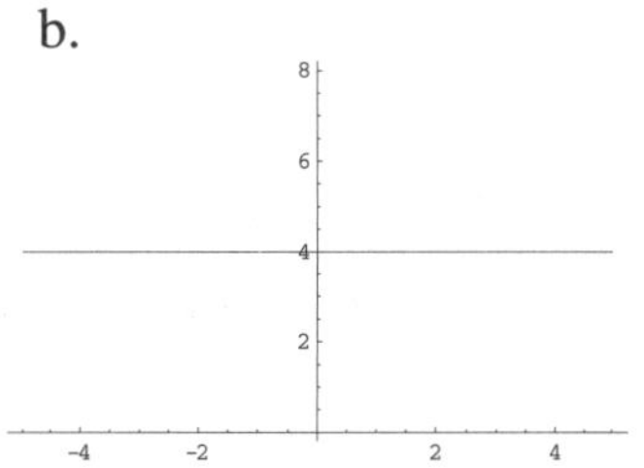

c.

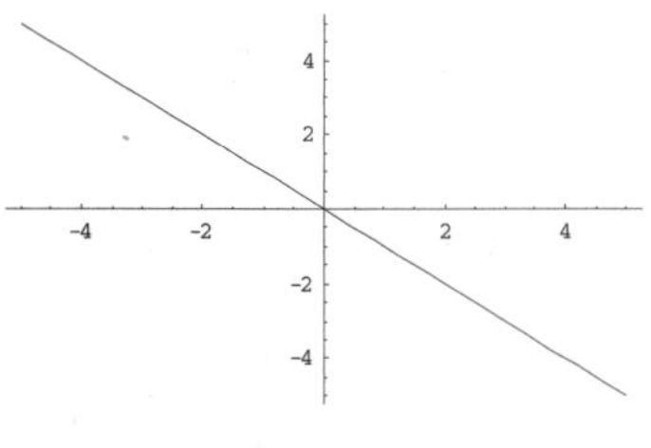

d.

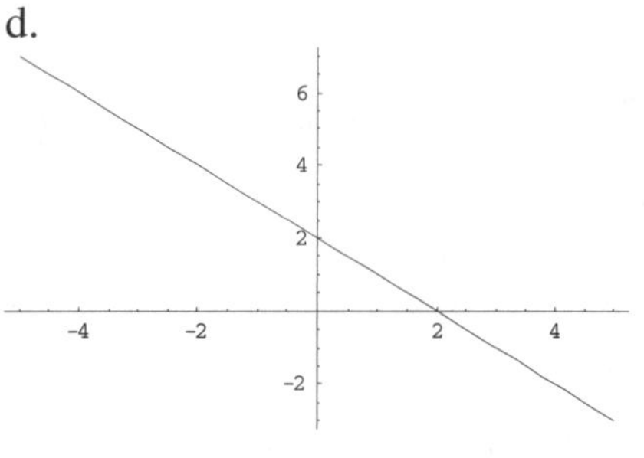

For Problems 12 – 23, give the slope of the line (if it exists) and the *y*-intercept (if it exists).

12. $y = 3x - 7$ 13. $4x - 2y = 9$ 14. $y = 11 - 5x$

15. $y = \frac{x}{6} - 2$ 16. $\frac{y}{5} + \frac{x}{3} = -1$ 17. $y = 4x$

18. $3y = x$ 19. $2y = 4x - 1$ 20. $x = 9$

21. $-y = 5x + 3$ 22. $y = 9$ 23. $60x - 12y = 36$

For the functions in Problems 24 – 26, do the following:

a. Find the slope and y-intercept of the graph of the function.
b. Determine if the graph is rising or falling.

24. $y = 2x - 1$; [-5, 5]

25. $y = 0.002x + 0.05$; [-100, 100]

26. $y = 20{,}000 - 100x$; [0, 500]

27. Rank the functions in Problems 24 – 26 in order of increasing steepness.

For each of the functions in Problems 28 – 33, find the rate of change.

28. $y = 7x - 2$ 29. $y = \frac{1}{5}x + 15$ 30. $y = 40 - 5x$

31. $y = -8x + 10$ 32. $y = 200x - 50$ 33. $y = 15 + 75x$

34. If a linear function has the points (4, 7) and (6, 10) on its graph, what is the rate of change of the function?

35. If a linear function has the points (-2, 8) and (4, 10) on its graph, what is the rate of change of the function?

36. If a linear function has the points (3, 7) and (-1, 12) on its graph, what is the rate of change of the function?

37. One line passes through the points (0, 4) and (2, 8), and a second line passes through the points (1, 3) and (3, 7). Are the lines parallel, perpendicular, or neither?

38. One line passes through the points (-2, 1) and (1, 4), and a second line has equation $x + y = 5$. Are the lines parallel, perpendicular, or neither?

39. One line passes through the points (0, 5) and (-2, 9), and a second line has equation $y = \frac{x}{2} + 4$. Are the lines parallel, perpendicular, or neither?

40. Graph the identity function. What is its rate of change?

41. Graph the constant function $f(x) = 2$. What is its rate of change?

Section 1.4 Equations of Lines

1. Write the equation of the line that has slope 3 and y-intercept -5.

2. Write the equation of the line that has slope -6 and y-intercept 4.

3. Write the equation of the line that has slope $\frac{2}{3}$ and y-intercept 0.

4. Write the equation of the line that has slope -1 and y-intercept $\frac{4}{5}$.

5. Write the equation of the line through the point (3, -2) having slope $m = 6$.

6. Write the equation of the line through the point (-2, 5) having slope $m = -8$.

7. Write the equation of the line through the point (-1, -4) having slope $m = 3$.

8. Write the equation of the line through the point (-6, 0) having slope $m = -2$.

9. Write the equation of the line through the point (4, 3) having slope $m = 2$.

10. Write the equation of the line that has slope -1 and passes through (-4, -1).

11. Write the equation of the line that has zero slope and passes through (3, 6).

12. Write the equation of the vertical line through the point (-5, 2).

13. Write the equation of the line with undefined slope and passes through (1, -3)

14. Write the equation of the horizontal line through (0, 7).

15. Write the equation of the line that passes through (-2, 1) and (5, 2).

16. Write the equation of the line that passes through (-3, -6) and (0, 2).

17. Write the equation of the line that passes through (-1, 1) and (4, 3).

18. Write the equation of the line that passes through (8, 2) and (5, 2).

19. Write the equation of the line that passes through (-2, 2) and (-1, -5).

20. Write the equation of the line that passes through (-3, -7) and (4, 6).

21. Write the equation of the line that passes through (1, 0) and (1, 7).

22. Write the equation of the line that passes through (-4, -3) and (-1, -5).

23. Write the equation of the line having x-intercept -3 and y-intercept 1.

24. Write the equation of the line having x-intercept 4 and y-intercept 7.

25. Write the equation of the line whose graph is shown.

a.

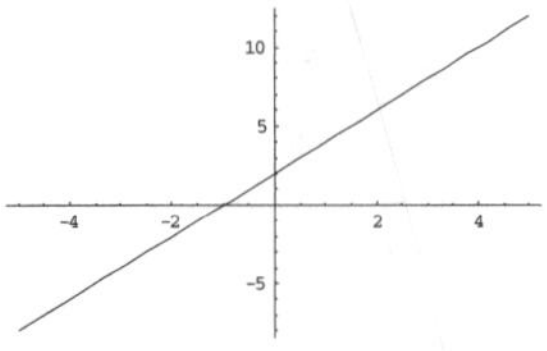

b.

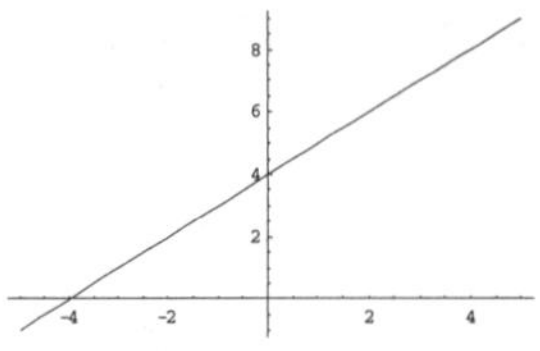

c.

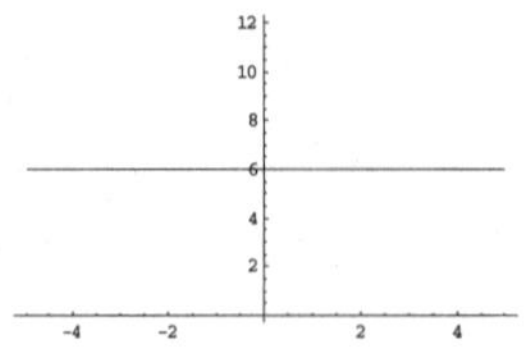

26. Write the equation of a function if its rate of change is -4 and $y = 6$ when $x = 0$.

27. Write the equation of a function if its rate of change is $\frac{1}{4}$ and $y = -2$ when $x = 0$.

28. For the function $y = x^2$, compute the average rate of change between $x = 4$ and $x = 7$.

29. For the function $y = x^2$, compute the average rate of change between $x = 2$ and $x = 3$.

For Problems 30 – 35, find $\dfrac{f(x+h)-f(x)}{h}$.

30. $f(x) = 3x - 5$.

31. $f(x) = -x + 4$

32. $f(x) = x^2 + 2$

33. $f(x) = 4x^2 - 6$

34. $f(x) = x^2 - 4x$

35. $f(x) = 3x^2$

In Problems 36 – 38, use each table that gives a set of input values x and the corresponding outputs y that satisfy a function.

36. Write the equation of the linear function that fits the data.

x	2	4	6	8	10
y	20	26	32	38	44

37. Write the equation of the linear function that fits the data.

x	5	10	15	20	25
y	100	130	160	190	220

38. Write the equation of the linear function that fits the data.

x	50	60	70	80	90
y	30	28	26	24	22

Section 2.1 Algebraic and Graphical Solutions of Linear Equations

In Problems 1 – 12, solve the following equations.

1. $2x-7=6x+9$
2. $3x+5=7-2x$
3. $4(x+1)=7+x$
4. $5(2x-4)=3x+8$
5. $6(2x+1)=3(x-6)$
6. $x+\frac{1}{2}=2x-5$
7. $2x+\frac{1}{4}=x-\frac{2}{3}$
8. $\frac{2(x+1)}{3}=x-2$
9. $\frac{4(x-5)}{5}-x=2x-15$
10. $3.4x=2.1x+5.2$
11. $\frac{2}{3}x+\frac{1}{2}=\frac{1}{3}x-\frac{1}{6}$
12. $\frac{1}{4}x-3=\frac{2}{5}x+\frac{3}{10}$

For Problems 13 – 15, find

a. the solution to the equation $f(x)=0$
b. the x-intercept of the graph of $y=f(x)$
c. the zero of $f(x)$

13. $f(x)=45+1.5x$
14. $f(x)=48-12x$
15. $f(x)=\frac{x+6}{3}$

In Problems 16 and 17, you are given the graph of a certain function $y=f(x)$ and the zero of the function. Using this graph, find

a. the x-intercept of the graph of $y=f(x)$
b. the zero of $f(x)$

16.

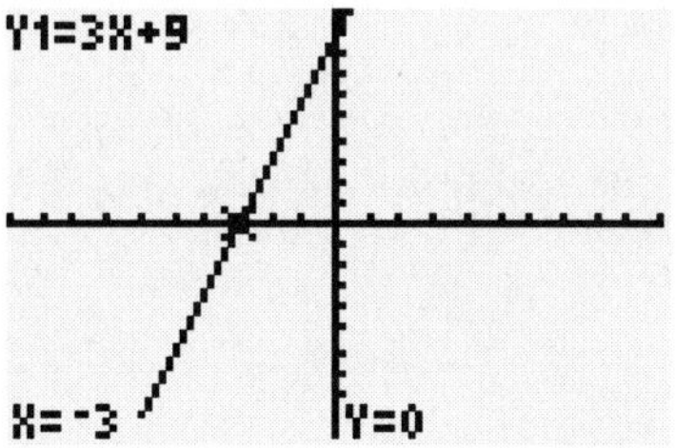

17.

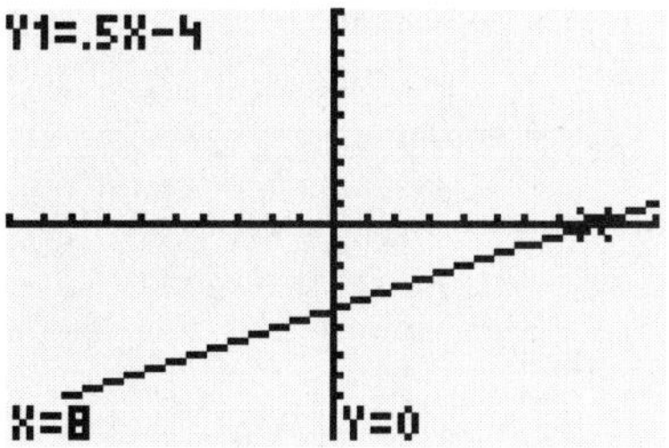

In Exercises 18 – 21, you are given a table showing input and output values for a given function $y = f(x)$. Using this table, find (if possible)

a. the x-intercept of the graph of $y = f(x)$

b. the y-intercept of the graph of $y = f(x)$

c. the solution to the equation $f(x) = 0$

18.

X	Y1
-1	6
0	4
1	2
2	0
3	-2
4	-4
5	-6

X= -1

19.

X	Y1
-1	-20
0	-15
1	-10
2	-5
3	0
4	5
5	10

X= -1

20.

X	Y1
-1	-1.5
0	-3
1	-4.5
2	-6
3	-7.5
4	-9
5	-10.5

X= -1

21.

X	Y1
-1	1.9
0	2
1	2.1
2	2.2
3	2.3
4	2.4
5	2.5

X= -1

22. Solve $t(2-y) = ax$ for y.

23. Solve $A = \frac{1}{2}h(b_1 + b_2)$ for h.

24. Solve $ax + by = cx + d$ for x.

25. Solve $3x + 2y = a(x+2)$ for x.

For Problems 26 – 29, solve the following equations for y and graph them on a standard window on a graphing utility.

26. $2x + y = -4$

27. $x^2 - y = 2$

28. $2y - 4x^2 = 9$

29. $x^3 - y = 0$

Section 2.2 Fitting Lines to Data Points: Modeling Linear Functions

For Problems 1 and 2, using the data represented in each graph below, state whether or not it represents a linear relationship.

1.

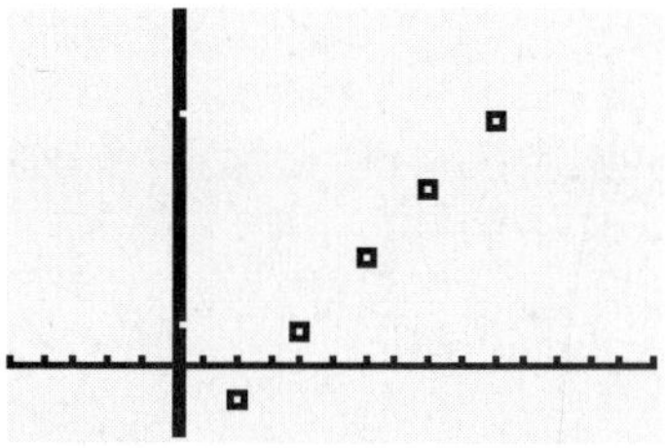

2.

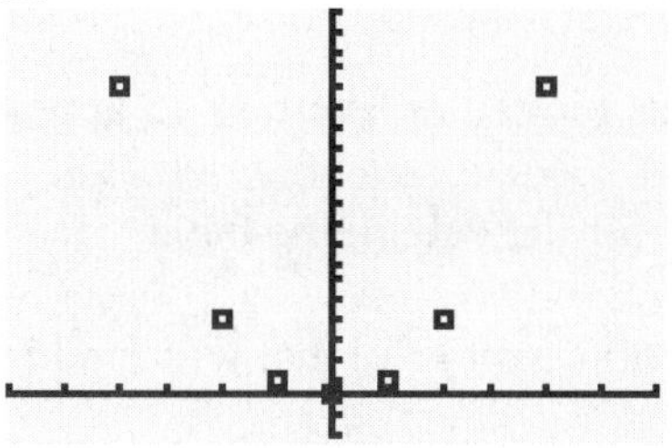

For Problems 3 – 14, use the three tables below.

Table A

x	y
1	-3
3	-1
5	1
7	3
9	5
11	7

Table B

x	y
-8	5
-4	2.5
0	0
4	-2.5
8	-5
12	-7.5

Table C

x	y
10	5
20	20
30	35
40	50
50	65
60	80

3. Create a scatter plot for the series of data in Table A.

4. Create a scatter plot for the series of data in Table B.

5. Create a scatter plot for the series of data in Table C.

6. Can the series of data in Table A be fit to a linear function? Why or why not?

7. Can the series of data in Table B be fit to a linear function? Why or why not?

8. Can the series of data in Table C be fit to a linear function? Why or why not?

9. Find the linear function $f(x)$ that is the best fit for the series of data in Table A.

10. Find the linear function $g(x)$ that is the best fit for the series of data in Table B.

11. Find the linear function $h(x)$ that is the best fit for the series of data in Table C.

12. For $f(x)$, find $f(10)$ and $f(-2.5)$.

13. For $g(x)$, find $g(15)$ and $g(-10)$.

14. For $h(x)$, find $h(1.5)$ and $h(-7)$.

Use the data in the table for Problems 15 – 18.

x	1	3	6	8	11	13	15
y	3	6	13	18	23	26	31

15. Construct a scatter plot of the data in the table.

16. Determine if the points plotted in Exercise 15 appear linear.

17. Create a linear model for the data in the table.

18. Use the rounded function $y = f(x)$ that was reported in Problem 17 to evaluate $f(7)$.

Use the data in the table for Problems 19 – 22.

x	10	15	20	25	30	35	40
y	38	62	82	100	115	143	161

19. Construct a scatter plot of the data in the table.

20. Determine if the points plotted in Problem 19 appear linear.

21. Create a linear model for the data in the table.

22. Use the rounded function $y = f(x)$ that was reported in Exercise 21 to evaluate $f(7)$.

23. Determine which of the equations $y = 3x + 1$ or $y = 2.5 + 1$, is the better fit for the data points (0, 1), (1, 4), (2, 6), (3, 10), (4, 12).

24. Determine which of the equations $y = 3x - 2.5$ or $y = 2.5x - 2$, is the better fit for the data points (0, -2.5), (1, .5), (2, 3), (3, 5), (4, 8).

25. Without graphing, determine which of the following data sets are exactly linear, approximately linear, or nonlinear.

a.

x	y
1	2
2	5
3	6
4	9
5	10

b.

x	y
1	1
2	4
3	10
4	16
5	25

c.

x	y
1	5
2	8
3	11
4	14
5	17

For Problems 26 – 30, match the data with the equation that gives the best fit.

26. (0, 0), (1, 4), (2, 9), (3, 11), (4, 16)	a. $y = 2x - 10$
27. (0, 1), (2, -4), (4, -10), (6, -15)	b. $y = 3.9x + 0.2$
28. (1, 5), (3, 1), (5, -4), (7, -9), (9, -13)	c. $y = -2.3x + 7.5$
29. (10, 10), (15, 20), (20, 25), (30, 33), (40, 40)	d. $y = -2.7x + 1.1$
30. (-1, 6), (-2, 3), (-3, -1), (-4, -4), (-5, -7)	e. $y = 3.3x + 9.3$

Section 2.3 Systems of Linear Equations in Two Variables

1. What are the coordinates of the point of intersection of $y = 2x + 3$ and $y = 3x + 2$?

2. What are the coordinates of the point of intersection of $y = x + 3$ and $y = 2x + 5$?

3. What are the coordinates of the point of intersection of $y = 4 - x$ and $y = 3x + 4$?

4. What are the coordinates of the point of intersection of $y = 4x + 2$ and $y = 6 + 4x$?

5. Give the coordinates of the point of intersection of $x + y = 7$ and $2x + y = 10$.

6. Give the coordinates of the point of intersection of $3x + y = -1$ and $x - y = 3$.

In Problems 7 – 12, solve the systems of equations graphically.

7. $\begin{cases} x + y = 3 \\ x - y = -5 \end{cases}$

8. $\begin{cases} 2x + y = -3 \\ x - 3y = 9 \end{cases}$

9. $\begin{cases} x + y = 7 \\ 2x - y = 11 \end{cases}$

10. $\begin{cases} 3x + 2y = 13 \\ x + 5y = 13 \end{cases}$

11. $\begin{cases} x + 3y = 2 \\ 3x - 2y = -16 \end{cases}$

12. $\begin{cases} 2x + y = -3 \\ 2x - 3y = -7 \end{cases}$

13. Does the system $\begin{cases} 2x + 5y = 8 \\ -4x - 10y = 7 \end{cases}$ have a unique solution, no solution, or many solutions? What does this mean graphically?

14. Does the system $\begin{cases} 2x + 5y = 8 \\ -4x - 10y = -16 \end{cases}$ have a unique solution, no solution, or many solutions? What does this mean graphically?

In Problems 15 – 18, solve the systems of equations by substitution.

15. $\begin{cases} x - 2y = 1 \\ 2x + y = 12 \end{cases}$

16. $\begin{cases} 2x - y = 10 \\ 3x + 2y = 5 \end{cases}$

17. $\begin{cases} x + y = 6 \\ 3x - 4y = -3 \end{cases}$

18. $\begin{cases} 3x + 2y = 14 \\ 2x + 3y = 6 \end{cases}$

In Problems 19 – 22, solve the systems of equations by elimination, if a solution exists.

19. $\begin{cases} x+y=-2 \\ 2x-y=17 \end{cases}$

20. $\begin{cases} 2x+y=6 \\ 4x-3y=-13 \end{cases}$

21. $\begin{cases} 1.5x+y=6 \\ 3x+4y=12 \end{cases}$

22. $\begin{cases} 3x-5y=4 \\ 6x-2y=-8 \end{cases}$

In Exercises 23 – 30, solve the systems of equations by any convenient method, if a solution exists.

23. $\begin{cases} y=3x-4 \\ 2x+y=1 \end{cases}$

24. $\begin{cases} y=2x+1 \\ 4x-2y=-2 \end{cases}$

25. $\begin{cases} 2x+3y=-4 \\ 6x-4y=-38 \end{cases}$

26. $\begin{cases} x+3y=3 \\ 3x+6y=10 \end{cases}$

27. $\begin{cases} 2.5x+4y=18 \\ 5x-1.5y=17 \end{cases}$

28. $\begin{cases} 2x-y=10 \\ 6x-3y=5 \end{cases}$

29. $\begin{cases} 3x+1.5y=9.5 \\ 6x-y=-1 \end{cases}$

30. $\begin{cases} 3x+2y=14 \\ -9x-6y=-42 \end{cases}$

Section 2.4 Solution of Linear Inequalities

For Problems 1 – 18, solve the following inequalities both algebraically and graphically.

1. $5x-7\le 2x+5$
2. $x-7>3x+2$
3. $6-4x<2x+18$
4. $3(5-x)\ge 6-2x$
5. $5(2x-2)<4(x+1)$
6. $3(7-2x)<6(x-2)$
7. $3x+4\ge \frac{x}{3}+12$
8. $5-\frac{2}{3}x>\frac{1}{6}+2x$
9. $3x-\frac{3}{4}>\frac{2}{5}x+1.85$
10. $5-2(x+4)\le 6x+4$
11. $3x+7>x+2(x-5)$
12. $4x+2(x-1)<3x-7$
13. $2.5x+1.3<3.6x-4.2$
14. $4.2x+3>2x-1.6$
15. $2.4x-1.2(2x+4)<3.4$
16. $\frac{3x+2}{5}\ge \frac{2x+7}{3}$
17. $\frac{3x-1}{4}>\frac{x-6}{2}$
18. $\frac{2(x-5)}{3}>\frac{3(2-x)}{2}$
19. Use the x-intercept method to solve the inequality $2x+8\le 4x-6$.
20. Use the x-intercept method to solve the inequality $3x-7<x+5$.
21. Solve the inequality $4x+2\le 6-2x$ by the intersection method.
22. Solve the inequality $x+2\le 6-3x$ by the intersection method.

In Problems 23 – 32, solve the double inequalities.

23. $3<x+6<10$
24. $2\le 5-x\le 8$
25. $10>3x-5>4$
26. $30<10x+20\le 60$
27. $x+5>1$ and $x+2>6$
28. $2x+3\ge 9$ and $2x+3<-3$
29. $x+5>2x+6$ or $3x+2<x-8$
30. $\frac{1}{2}x-4\ge 3$ or $3x+2\ge 8$
31. $5<x+2<7$ or $2x+4>12$
32. $3x+6<9$ or $5>\frac{1}{3}x+1>1$

Section 3.1 Quadratic Functions; Parabolas

In Problems 1 – 6, (a) determine if the function is quadratic. If it is (b) determine if the graph is concave up or concave down. (c) Determine if the vertex of the graph is a maximum point or a minimum point.

1. $y = 3x^2 - 6x$
2. $y = 5x - x^2$
3. $y = \dfrac{1}{x^2 + 2x - 5}$
4. $y = 2x^2 + 4x - 7$
5. $y = 2x^3 + 9x^2 + 6x - 5$
6. $y = 0.3 - 0.4x + 0.01x^2$

In Problems 7 – 14 (a) graph each quadratic function on [10, -10] by [10, -10]. (b) Does this window give a complete graph?

7. $y = 2x^2 + 6x - 3$
8. $y = x^2 + 8x - 4$
9. $y = -x^2 - 10x - 4$
10. $y = .1x^2 - 10x - 4$
11. $y = 1 - x^2$
12. $y = 20x^2 + 5$
13. $y = .01x^2 - 6x$
14. $y = .1x^2 + 4x$

In Problems 15 – 24, (a) give the coordinates of the vertex of the graph of each function. (b) Graph each function on a window that includes the vertex.

15. $y = (x-2)^2 + 3$
16. $y = (x+1)^2 - 3$
17. $y = 6x - 3x^2$
18. $y = 3(x+2)^2 - 2$
19. $y = -2(x+4)^2 + 2$
20. $y = -(x+3) - 4$
21. $y = x^2 + 7$
22. $y = 2x^2 + 4x$
23. $y = 3x^2 - 6x + 2$

In Problems 24 – 26, sketch complete graphs of the functions.

24. $y = x^2 + 16x + 64$
25. $y = -x^2 + 50x + 600$
26. $y = 2x^2 - 35x - 750$

Use the graph of each function in Problems 29 – 32 to estimate the x-intercepts.

27. $y = x^2 + 9x - 90$
28. $y = 2x^2 + x - 15$
29. $y = -5x^2 + 36x + 32$

Section 3.2 Solving Quadratic Equations

In Problems 1 – 9, solve each equation algebraically.

1. $x^2+2x-3=0$
2. $x^2-x-20=0$
3. $x^2+4x-12=0$
4. $2x^2-7x-15=0$
5. $6x^2+29x+9=0$
6. $6x^2+5x-4=0$
7. $5x^2+42x+16=0$
8. $4x^2+12x+9=0$
9. $6x^2-23x=4$

In Problems 10 – 15, use a graphing utility to find or approximate the x-intercept of the graph of each function.

10. $y=x^2+2x-8$
11. $y=x^2+13x+30$
12. $y=2x^2-13x+6$
13. $y=6x^2-13x+5$
14. $y=8x^2-6x-9$
15. $y=6x^2-26x+8$

Use a graphing utility as an aid in factoring to solve the equations in Problems 16 – 21.

16. $x^2-10x-1200=0$
17. $x^2+55x+600=0$
18. $2x^2+85x=750$
19. $2x^2-75x=200$
20. $200x^2-45=90x$
21. $5x^2+62x+120=0$

In Problems 22 – 24, use the square root method to solve the quadratic equations.

22. $2x^2-18=0$
23. $x^2-28=0$
24. $5x^2-60=0$

In Problems 25 – 27, complete the square to solve the quadratic equations.

25. $x^2-8x+12=0$
26. $2x^2+11x+12=0$
27. $x^2-3x-70=0$

In Problems 28 – 33, use the quadratic formula to solve the equations.

28. $6x^2-5x-4=0$
29. $6x^2+4x-1=0$
30. $2x^2+19=14x$
31. $x^2+18=10x$
32. $2x^2-2x=1$
33. $2x^2+3x-4=0$

In Problems 34 – 39, find the exact solution to $f(x)=0$ in the complex numbers and confirm that the solutions are not real by showing that the graph of $y=f(x)$ does not cross the x-axis.

34. $x^2=4x-5$
35. $x^2+35=0$
36. $x^2-6x+14=0$
37. $x^2+4x+11=0$
38. $4x^2-4x=-7$
39. $9x^2-6x+5=0$

Section 3.3 Piecewise-Defined Functions and Powers Functions

In Problems 1 – 10, sketch the graph of each function using the window [-5, 5] by [-5, 5].

1. $y = -2x^3$

2. $y = 3x^4$

3. $y = -x^{1/3}$

4. $y = \sqrt{x} - 1$

5. $y = \sqrt[3]{x} + 2$

6. $y = \dfrac{x-1}{x}$

7. $y = \dfrac{3}{x}$

8. $y = \begin{cases} 2, & x < 0 \\ 4, & x \geq 0 \end{cases}$

9. $y = \begin{cases} 3, & x \leq 1 \\ 1, & x > 1 \end{cases}$

10. $y = \begin{cases} -2, & x < -1 \\ 2, & x \geq -1 \end{cases}$

11. a. Graph $f(x) = \begin{cases} -2x+1, & x \leq 2 \\ x+1, & x > 2 \end{cases}$ b. Find $f(1)$ and $f(5)$.

12. a. Graph $f(x) = \begin{cases} 2-x, & x \leq 0 \\ x^2, & x > 0 \end{cases}$ b. Find $f(-2)$ and $f(3)$.

13. a. Graph $f(x) = |x+3|$ b. Find $f(-4)$ and $f(6)$.

14. a. Graph $f(x) = |2-x|$ b. Find $f(4)$ and $f(8)$.

For each of the functions in Problems 15 – 23, find the value of $f(-1)$ and $f(2)$.

15. $f(x) = 2(x-1)^3$

16. $f(x) = 2x^2 + 5$

17. $f(x) = \sqrt{x-2}$

18. $f(x) = \sqrt[3]{x}$

19. $f(x) = \dfrac{1}{x+1}$

20. $f(x) = \dfrac{1}{2x} + 2$

21. $y = \begin{cases} 4, & x < 1 \\ 8, & x \geq 1 \end{cases}$

22. $y = \begin{cases} x+4, & x \leq 0 \\ 2x, & x \geq 0 \end{cases}$

23. $y = \begin{cases} 2x^2+1, & x < 2 \\ 8-x^2, & x \geq 2 \end{cases}$

24. Determine if the function $y = -2x^3$ is increasing or decreasing for a. $x < 0$
b. $x > 0$

Section 3.4 Quadratic and Power Models

For Problems 1 – 5, use 2nd differences to determine if the data is exactly quadratic, approximately quadratic, or not quadratic.

1.

x	-2	-1	0	1	2	3
$f(x)$	3	0	-1	0	3	8

2.

x	-2	-1	0	1	2	3
$f(x)$	-4	2	2	2	8	26

3.

x	-2	3	8	13	18	23
$f(x)$	-2	18	88	208	378	598

4.

x	-2	3	8	13	18	23
$f(x)$	1	-4	-60	-160	-310	-530

5.

x	2	4	6	8	10	12
$f(x)$	6	16	32	63	91	140

6. Find the quadratic function that models the data in the table below.

x	-3	-2	-1	0	1	2	3	4
y	5	-2	-5	-4	1	10	23	40

7. Find the quadratic function that models the data in the table below.

x	2	4	6	8	10	12	14
y	5	-1	-11	-31	-59	-95	-139

8. a. Make a scatter plot of the data in the table below.

x	1	2	3	4	5	6	7	8
y	6.4	7.51	8.23	8.8	9.27	9.66	10.01	10.33

   b. Find a power function that models the data.
   c. Find a linear function that models the data.
   d. Visually determine which model is the better fit for the data.

9. Find a power function that models the data in the table below.

x	0	2	4	6	8	10
y	0	2.30	5.28	8.59	12.13	15.85

10. Find a power function that models the data in the table below.

x	1	2	3	4	5	6	7
y	0.1	0.57	1.56	3.20	5.59	8.82	12.96

For Problems 11 – 13, use the table below.

x	-2	-1	0	1	2	3	4	5
$f(x)$	0.7	0.2	0.1	0.4	1.1	2.2	3.7	5.6

11. Create a scatter plot for the data.

12. Find both a linear and a quadratic model for the set of data.

13. Which is the better fit, the linear or the quadratic model?

For Problems 14 – 16, use the table below.

x	100	128	170	195	200	212	224	237
$f(x)$	51450	65786	87290	100090	102650	108794	114938	115450

14. Create a scatter plot for the data.

15. Find both a linear and a quadratic model for the set of data.

16. Which is the better fit, the linear or the quadratic model?

Section 4.1 Transformations of Graphs and Symmetry

In Problems 1 – 10, sketch the graph of each pair of functions using a standard window.

1. $y = x^3$, $y = 3 + x^3$
2. $y = x^2$, $y = x^2 + 1$
3. $y = x^2$, $y = (x-6)^2$
4. $y = \sqrt[3]{x}$, $y = \sqrt[3]{x+1} - 3$
5. $y = x^3$, $y = (x+2)^3 - 5$
6. $y = |x|$, $y = -|x-6|$
7. $y = |x|$, $y = |x| - 2$
8. $y = \frac{1}{x}$, $y = \frac{4}{x}$
9. $y = x^2$, $y = -0.2x^2$
10. $y = \sqrt{x}$, $y = \frac{1}{4}\sqrt{x}$

11. How is the graph of $y = (x+1)^2 - 3$ transformed from the graph $y = x^2$?

12. How is the graph of $y = (x-4)^3 + 6$ transformed from the graph $y = x^3$?

13. Suppose the graph of $y = \sqrt{x}$ is shifted up 2 units and to the left 4 units. What is the equation that gives the new graph?

14. Suppose the graph of $y = x^{1/3}$ is shifted down 3 units and to the right 1 unit. What is the equation that gives the new graph?

In Problems 15 – 20, determine algebraically whether the graph of the given equation is symmetric with respect to the *x*-axis, *y*-axis, and/or the origin.

15. $y = 4x^2 + 1$
16. $y = -x^2 + 6$
17. $y = x^3$
18. $x^2 + y^2 = 4$
19. $x^2 - y^2 = 9$
20. $y = -x^3 + 5x$

In Problems 21 – 23, determine whether the function is even, odd, or neither.

21. $y = |x| + 2$
22. $y = 2x^3 - 3x$
23. $y = x^4 + 2x^2$

Section 4.2 Combining Functions; Composite Functions

In Problems 1 – 6, find (a) $(f+g)(x)$ (b) $(f-g)(x)$ (c) $(f\cdot g)(x)$ (d) $\left(\dfrac{f}{g}\right)(x)$ (e) the domain of $\left(\dfrac{f}{g}\right)(x)$.

1. $f(x)=2x+7$ $\quad g(x)=x-4$

2. $f(x)=x^2+6$ $\quad g(x)=x^2-4$

3. $f(x)=x^2+2x$ $\quad g(x)=3x-5$

4. $f(x)=\dfrac{1}{x}$ $\quad g(x)=\dfrac{4}{x+2}$

5. $f(x)=\sqrt{x}$ $\quad g(x)=x^2-6x$

6. $f(x)=\dfrac{x+2}{3}$ $\quad g(x)=\dfrac{2x-1}{6}$

7. If $f(x)=x^3+3x$ and $g(x)=4-x^2$, evaluate each of the following:
   a. $(f+g)(2)$
   b. $(f-g)(-1)$
   c. $(f\cdot g)(3)$
   d. $\left(\dfrac{f}{g}\right)(-4)$

8. If $f(x)=\sqrt{x+1}$ and $g(x)=\dfrac{2}{x+1}$, evaluate each of the following:
   a. $(f+g)(3)$
   b. $(f-g)(0)$
   c. $(f\cdot g)(8)$
   d. $\left(\dfrac{f}{g}\right)(15)$

In Problems 9 – 14, find (a) $(f\circ g)(x)$ and (b) $(g\circ f)(x)$.

9. $f(x)=x+1$ $\quad g(x)=3x+8$

10. $f(x)=6-x$ $\quad g(x)=7-2x$

11. $f(x) = 4x^2$ $\quad g(x) = x + 9$

12. $f(x) = \dfrac{1}{x}$ $\quad g(x) = 6x^2$

13. $f(x) = |x + 3|$ $\quad g(x) = 4x + 2$

14. $f(x) = \sqrt{x - 5}$ $\quad g(x) = 3x^2$

15. Let $f(x) = 3x + 4$ and $g(x) = 2x^2 - 1$. Evaluate each expression.
    a. $(f \circ g)(1)$
    b. $(f \circ g)(-3)$
    c. $(g \circ f)\left(\dfrac{1}{3}\right)$
    d. $(g \circ f)(-4)$

16. Use the graphs of $f(x)$ and $g(x)$ to evaluate each expression.

$f(x)$

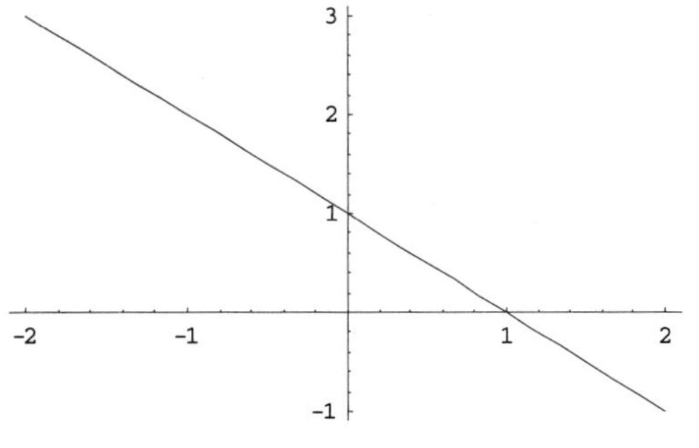

$g(x)$

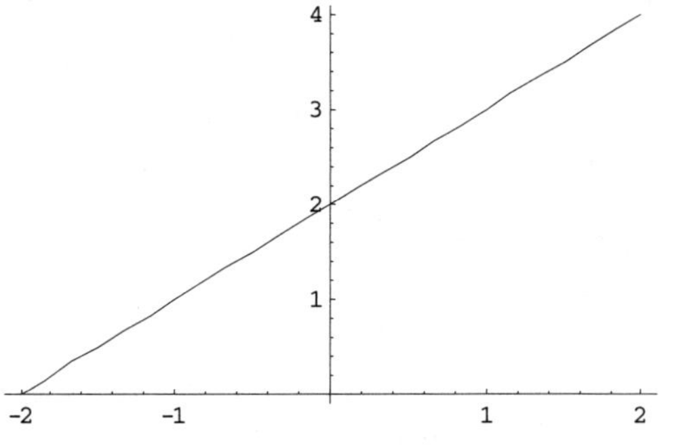

a. $(f + g)(1)$ b. $(f \cdot g)(2)$ c. $\left(\dfrac{f}{g}\right)(-1)$

d. $(f \circ g)(0)$ e. $(g \circ f)(-1)$ f. $(g \circ f)(1)$

Section 4.3 Inverse Functions

In Problems 1 – 4, determine if the function f defined by the set of ordered pairs has an inverse. If so, find the inverse.

1. {(1, 4), (2, 8), (3, 6), (4, 10)}

2. {(3. 1), (3. 2). (3. 3), (3, 4)}

3. {(4, 4), (5, 5), (6, 6)}

4. {(-1, 0), (1, 2), (2, 3)}

5. Let $f(x)=\frac{6}{x}$ and $g(x)=6x$.
   a. Find $f(g(x))$ and $g(f(x))$.
   b. Are $f(x)$ and $g(x)$ inverse functions?

6. Let $f(x)=2x-3$ and $g(x)=\frac{x+3}{2}$.
   a. Find $f(g(x))$ and $g(f(x))$.
   b. Are $f(x)$ and $g(x)$ inverse functions?

7. Let $f(x)=\frac{x}{3}-4$ and $g(x)=3x-4$.
   a. Find $f(g(x))$ and $g(f(x))$.
   b. Are $f(x)$ and $g(x)$ inverse functions?

8. Let $f(x)=\sqrt{x+5}$ and $g(x)=(x-5)^2$.
   a. Find $f(g(x))$ and $g(f(x))$.
   b. Are $f(x)$ and $g(x)$ inverse functions?

9. Let $f(x)=\sqrt[3]{x-2}$ and $g(x)=x^3+2$.
   a. Find $f(g(x))$ and $g(f(x))$.
   b. Are $f(x)$ and $g(x)$ inverse functions?

10. For the function f described by $f(x)=5-2x$, complete the table below for f and f^{-1}.

x	$f(x)$
-2	9
-1	
0	
1	

x	$f^{-1}(x)$
9	-2
7	
5	
3	

11. For the function f described by $f(x)=\frac{1}{2}x^3+1$, complete the table below for f and f^{-1}.

x	$f(x)$
-2	-3
-1	
0	
1	

x	$f^{-1}(x)$
-3	-2
0.5	
1	
1.5	

12. Find the inverse of $f(x)=6x-3$.

13. Find the inverse of $f(x)=4x^3-3$.

14. Find the inverse of $f(x)=\sqrt{x+3}$.

15. For the function $f(x)=5x+2$, find $f^{-1}(x)$. Graph $f(x)$, $f^{-1}(x)$, and $y=x$ on the same axes.

16. For the function $f(x)=3x^3$, find $f^{-1}(x)$. Graph $f(x)$, $f^{-1}(x)$, and $y=x$ on the same axes.

17. For the function $f(x)=6+\sqrt{x}$, find $f^{-1}(x)$. Graph $f(x)$, $f^{-1}(x)$, and $y=x$ on the same axes.

For Problems 18 – 21, the graph of $y=f(x)$ is provided. Sketch the graph of the inverse function on the same set of axes.

18.

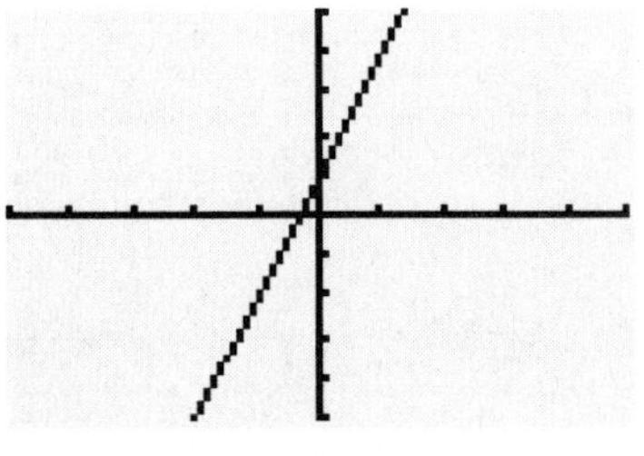

19.

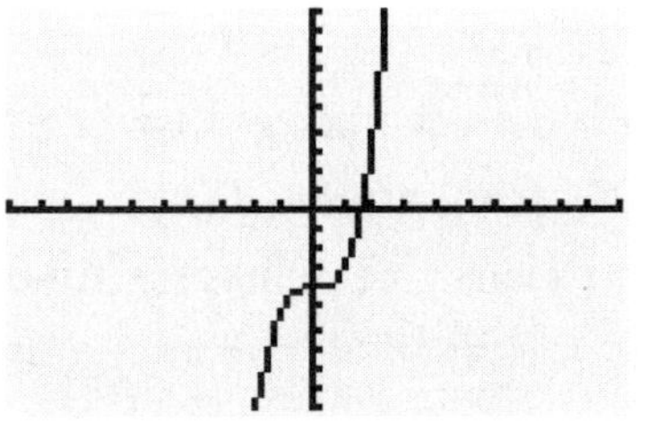

20.

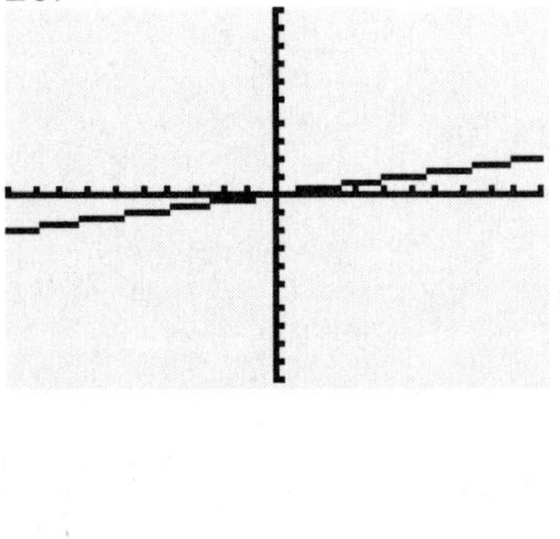

21.

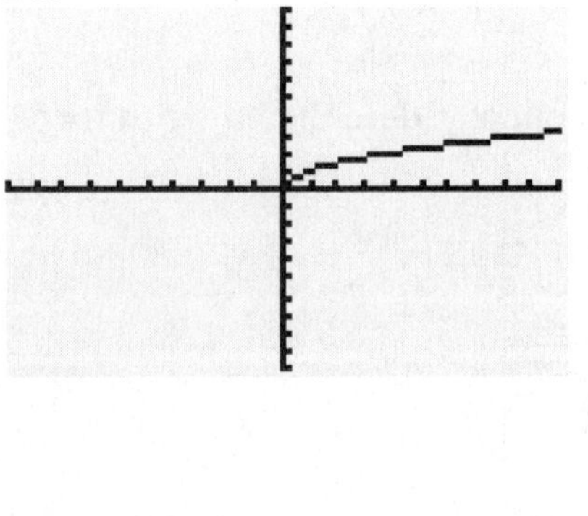

Section 4.4 Additional Equations and Inequalities

In Problems 1 – 10, solve the following equations and check graphically or by substitution.

1. $\sqrt{3x^2+9}-2x=0$
2. $\sqrt{x+3}=5$
3. $\sqrt{4x+1}-1=x$
4. $\sqrt[3]{2x}+5=3$
5. $\sqrt{x+2}=x$
6. $5+\sqrt{x+7}=x$
7. $\sqrt{x-2}+x=4$
8. $3\sqrt{x}-5=13$
9. $\sqrt{x+6}-\sqrt{x}=\sqrt{6}$
10. $\sqrt[3]{2x+6}=3$

In Problems 11 – 16, solve the equations and check graphically.

11. $|4x-2|=5$
12. $|x-6|=2$
13. $|3x-1=2x|$
14. $|x^2-9|=0$
15. $|x|=x^2+2x$
16. $|x+6|=x^2-4x$

In Problems 17 – 22, use analytical methods to solve the inequalities.

17. $x^2+2x<0$
18. $5x-x^2\le 0$
19. $x^2-3x-4\le 0$
20. $2x^2-4x-30<0$
21. $-x^2<9x+18$
22. $2x^2+3x>2$

In Problems 23 – 26, use graphical methods to solve the inequalities.

23. $x^2-4\le 7x$
24. $x^2-8x+5\le 0$
25. $x^2\ge x+1$
26. $x^2+2x-5>-x$

In Problems 27 – 30, solve the inequality by using analytical and graphical methods.

27. $(x+2)^3<1$
28. $(x-1)^4>16$
29. $|x-5|\ge 2$
30. $|x+4|<8$

Section 5.1 Exponential Functions

1. Which of the following are exponential functions?

a. $y = 4x^2 + 3$ b. $y = 5^x$ c. $y = x^{-e}$

d. $y = e^{-x}$ e. $y = \pi^{2x}$ f. $y = 3e$

2. Determine if each of the following functions is a growth exponential or a decay exponential.

a. $y = 3^{0.2x}$ b. $y = e^{-2x}$ c. $y = e^{0.3x}$

d. $y = \dfrac{1}{e^x}$ e. $y = (0.4)^{2x}$ f. $y = (1.2)^x$

3. a. Graph the function $f(x) = 3^x$ on the window [-5, 5] x [0, 100].
   b. Find $f(1)$, $f(-1)$ and $f(4)$, rounding answers to 3 decimal places.
   c. What is the horizontal asymptote of the graph?
   d. What is the *y*-intercept of the graph?

4. a. Graph the function $f(x) = e^{2x}$ on the window [-5, 5] x [0, 100].
   b. Find $f(1)$, $f(-1)$ and $f(4)$, rounding answers to 3 decimal places.
   c. What is the horizontal asymptote of the graph?
   d. What is the *y*-intercept of the graph?

In Problems 5 – 10, graph each function.

5. $y = 2^x$ 6. $y = 4^x - 3$ 7. $y = e^{x+1}$

8. $y = 2^{-x} + 2$ 9. $y = e^{2x}$ 10. $y = 5e^x + 3$

In Problems 11 – 16, use your knowledge of transformations to compare the graph of the functions with the graph of $f(x) = 6^x$. Then graph each function for $-5 \le x \le 5$.

11. $f(x) = 6^x - 1$ 12. $f(x) = 6^{-x}$ 13. $f(x) = 6^{x+2}$

14. $f(x) = 6^x + 5$ 15. $f(x) = 6^{2x}$ 16. $f(x) = 6^{-x} - 3$

In Problems 17 – 20, find the horizontal asymptote and the *y*-intercept for each function.

17. $y = e^x + 7$ 18. $y = 4(2^x) - 1$

19. $y = e^{-x} - 2$ 20. $y = e^{x-4}$

Section 5.2 Logarithmic Functions; Properties of Logarithms

In Problems 1 – 4, write the logarithmic equations in exponential form.

1. $y = \log_2 x$
2. $y = \log 5$
3. $y = \ln(3x)$
4. $y = \log_a w$

In Problems 5 – 8, write the exponential equations in logarithmic form.

5. $x = 3^b$
6. $y = e^{3x}$
7. $10^3 = y$
8. $8^{3x} = y$

In Problems 9 – 12, evaluate the logarithms, if possible. Round each answer to 3 decimal places.

9. $\log(23)$
10. $\ln(69,200)$
11. $\log(-50)$
12. $\log(100)$

In Problems 13 – 22, find the value of each logarithm without using a calculator.

13. $\log_2 8$
14. $\log 1000$
15. $\log_4 64$
16. $\log_3 \frac{1}{9}$
17. $\log 1$
18. $\ln e^2$
19. $\log_2 16$
20. $\log_4 16$
21. $\log_{16} 16$
22. $\log_9 3$
23. $\log 0.01$
24. $\log_4 32$

In Problems 25 – 28, graph the functions using technology.

25. $y = \ln x - 4$
26. $y = -2\ln x$
27. $y = \ln(x+2)$
28. $y = \ln(x-1)+5$

In Problems 29 – 32, rewrite each expression as the sum, difference, or product of logarithms, and simplify if possible.

29. $\ln \frac{3x+1}{2x-4}$
30. $\log\left[x^4(2x+1)^3\right]$

31. $\ln \frac{\sqrt[5]{4x+7}}{3x^2}$

32. $\log \sqrt{\frac{2x+1}{x^2+5}}$

In Problems 33 – 36, rewrite each expression as a single logarithm.

33. $6\ln x - 5\ln y$

34. $\frac{1}{4}\log(x+6x^3)$

35. $5\log(6x) + 2\log y - \log z$

36. $\frac{1}{5}(\ln x + 2\ln y)$

Section 5.3 Exponential and Logarithmic Equations

In Problems 1 – 6, solve the equations analytically and check graphically. Round answers to 3 decimal places.

1. $5200 = 10^x$
2. $3600 = 10^{3x}$
3. $31.5 = e^x$
4. $120 = e^{3x}$
5. $2000 = 50(10^x)$
6. $8000 = 200e^x$

In Problems 7 – 10, use the properties of logarithms to evaluate the expressions.

7. $\log 10^{13}$
8. $\ln e^7$
9. $2^{\log_2 5}$
10. $e^{\ln 2}$

In Problems 11 – 14, use $\log_a 12 = 1.079$ and $\log_a 4 = 0.602$ to evaluate each expression.

11. $\log_a 48$
12. $\log_a 3$
13. $\log_a 16$
14. $\log_a \sqrt{12}$

In Problems 15 – 18, use the change of base formula to evaluate each logarithm. Round answers to 3 decimal places.

15. $\log_5 8$
16. $\log_2 150$
17. $\log_4 \sqrt{6}$
18. $\log_6 \sqrt[4]{8}$

In Problems 19 – 32, solve the equations.

19. $\log_3 x = 4$
20. $\log_{16} x = -\frac{1}{4}$
21. $1 + 3\ln x = 7$
22. $5 + 4\log x = 17$
23. $16^x = 128$
24. $27^x = 2187$
25. $3(5^{2x}) = 1875$
26. $6^x = 30$
27. $4^{x-6} = 120$
28. $8(3^{2x}) = 340$

29. $\ln(3x+6)=9$

30. $\frac{1}{3}\ln(x+2)=4$

31. $\ln(x+6)-\ln(x)=3$

32. $2\ln x=\ln(7x-10)$

In Problems 33 and 34, solve the inequalities.

33. $4^x \le 180$

34. $5(3^x)>2160$

Section 5.4 Exponential and Logarithmic Models

1. Find the exponential function that models the data in the table below.

x	-2	-1	0	1	2	3	4	5
y	3/25	3/5	3	15	75	375	1875	9375

2. The following table has inputs x and the outputs for three functions: $f(x)$, $g(x)$, and $g(x)$. Which functions are exponential functions?

x	$f(x)$	$g(x)$	$h(x)$
1	2.5	12	17
2	5	18	24.5
3	7.5	27	85.75
4	10	40.5	300.125
5	12.5	60.75	1050.438
6	15	91.125	3676.531

3. Find the exponential function that models the data in the table below.

x	1	2	3	4	5	6
y	4.18	4.27	4.35	4.44	4.53	4.62

4. Find the exponential function that models the data in the table below.

x	1	2	3	4	5	6
y	200	325	527	854	1383	2241

5. a. Make a scatterplot of the data in the table below.

x	-2	-1	0	1	2
y	500	100	25	5	1

b. Does the data represent a linear function or an exponential function?
c. Find the best fit model for the data.

6. a. Make a scatterplot of the data in the table below.

x	-2	-1	0	1	2
y	5	2.5	.5	-1.5	-4

b. Does the data represent a linear function or an exponential function?
c. Find the best fit model for the data.

7. Find the logarithmic function that models the data in the table below.

x	1	2	3	4	5	6	7	8	9	10
y	3.6	4.10	4.25	4.40	4.50	4.65	4.75	4.80	4.90	4.95

8. Find the logarithmic function that models the data in the table below.

x	1	3	5	7	9
y	11	20	23	26	27

Section 5.5 Exponential Functions and Investing

In Problems 1 – 10, evaluate each expression. Round answers to 2 decimal places.

1. $20{,}000e^{0.04(10)}$

2. $6500e^{0.03(12)}$

3. $4000(1.05)^x$ for $x = 40$

4. $50000(1.08)^x$ for $x = 30$

5. $10{,}000\left(1+\dfrac{0.06}{k}\right)^{kn}$ for $k = 6,\ n = 15$

6. $93{,}000\left(1+\dfrac{0.12}{k}\right)^{kn}$ for $k = 4,\ n = 20$

7. $P\left(1+\dfrac{r}{k}\right)^{kn}$ for $P = 4000,\ r = 6\%,\ k = 12,\ n = 12$

8. $P\left(1+\dfrac{r}{k}\right)^{kn}$ for $P = 10000,\ r = 8\%,\ k = 4,\ n = 20$

9. $400\left[\dfrac{(1.06)^n - 1}{0.06}\right]$ for $n = 120$

10. $6000\left[\dfrac{(1.15)^n - 1}{0.15}\right]$ for $n = 60$

11. Find $f(1),\ f(2),\ f(3),\ f(3.5),\ f(4.5)$ if

$$f(x) = \begin{cases} 600 & 0 \le x < 1 \\ 800 & 1 \le x < 2 \\ 1000 & 2 \le x < 3 \\ 1200 & 3 \le x < 4 \\ 1400 & 4 \le x < 5 \end{cases}$$

12. Solve $A = Pe^{rn}$ for P.

13. Will deposits \$3600 in an account that compounds monthly at an annual rate of 9.2%. How much money will Will have in the account after 10 years?

14. Bethany deposits \$3600 in an account that compounds continuously at an annual rate of 9.2%. How much money will Bethany have in the account after 10 years?

Section 3.6 Annuities; Loan Repayments

In Problems 1 – 6, evaluate.

1. $2000\left[\dfrac{1-(1+0.01)^{-240}}{0.01}\right]$

2. $4200\left[\dfrac{1-(1+0.005)^{-120}}{0.005}\right]$

3. $160200\left[\dfrac{1-(1+0.045)^{-60}}{0.045}\right]$

4. $3000\left[\dfrac{0.05}{1-(1.05)^{-96}}\right]$

5. $900\left[\dfrac{0.12}{1-(1.12)^{-240}}\right]$

6. $30000\left[\dfrac{0.15}{1-(1.15)^{-120}}\right]$

6. What lump sum will grow to $20,000 in 6 years if it is invested at 8% compounded annually?

7. Find the present value of an annuity that will pay $500 at the end of each year for 10 years if the interest rate is 8% compounded annually.

8. The Nagle family is purchasing a new $450,000 home. They put a down-payment of $100,000 towards the purchase of the home and have a 30-year mortgage at 6% per year for the unpaid balance.
   a. What is the monthly payment for the Nagle family?
   b. What is the total amount the Nagle family will pay before they own the house outright?
   c. How much interest will the Nagle family pay over the life of the loan?

Section 5.7 Logistic and Gompertz Functions

In Problems 1 – 8, evaluate each expression.

1. $\dfrac{786}{1+1464^{-1.26(10)}}$

2. $\dfrac{205}{1+0.03e^{0.69(20)}}$

3. $\dfrac{100}{1+9.8e^{-0.98(x)}}$, where $x=30$

4. $\dfrac{1925}{1+321.1e^{1.86(x)}}$, where $x=50$

5. $2000(0.08)^{0.3^t}$, where $t=2$

6. $3000(0.04)^{0.5^t}$, where $t=1$

7. $1200(0.07)^{0.04^t}$, where $t=10$

8. $1000(0.06)^{0.03^t}$, where $t=20$

For Problems 9 – 12, use the function $f(x)=\dfrac{700}{1+4e^{-x}}$.

9. Graph the function from $x = 0$ to $x = 20$.

10. Find $f(0)$ and $f(10)$.

11. Is this function increasing or decreasing?

12. What is the limiting value of the function?

For Problems 13 -15, use the function $f(x)=200(0.12)^{0.5^t}$.

13. Graph the function from $x = 0$ to $x = 10$.

14. What is the initial value of this function?

15. What is the limiting value of this function?

Section 6.1 Higher-Degree Polynomial Functions

1. Graph the function $f(x) = 4x^3 + 6x^2 - 6x + 8$ on the windows given in (a), (b), (c) and (d). Which window gives a complete graph?
   a. [-10, 10] by [-10, 10]
   b. [-20, 20] by [-10, 10]
   c. [-2, 2] by [-5, 5]
   d. [-5, 5] by [-20, 20]

2. Graph the function $f(x) = 4x^3 - 2x^2 - 6x$ on the windows given in (a), (b), (c) and (d). Which window gives a complete graph?
   a. [-100, 100] by [-100, 100]
   b. [-2, 2] by [-2, 2]
   c. [-2, 2] by [-5, 5]
   d. [-5, 5] by [-2, 2]

3. Graph the function $f(x) = 4x^4 - x^2$ on the windows given in (a), (b), (c) and (d). Which window gives a complete graph?
   a. [-10, 10] by [-10, 10]
   b. [-2, 2] by [-2, 2]
   c. [-2, 2] by [0, 1]
   d. [-5, 5] by [-20, 20]

4. Graph the function $f(x) = -x^4 + 100x^2$ on the windows given in (a), (b), (c) and (d). Which window gives a complete graph?
   a. [-10, 10] by [-10, 10]
   b. [-10, 100] by [-10, 100]
   c. [-10, 1000] by [-10, 1000]
   d. [-10, 10000] by [-10, 10000]

For Problems 5 – 8, use the complete graph of the polynomial function to
   a. estimate the intercepts
   b. state whether the leading coefficient is positive or negative
   c. determine the minimum degree of the polynomial function

5.

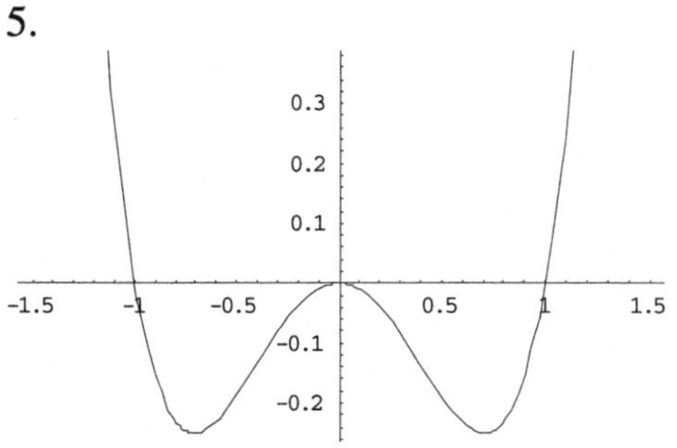

6.

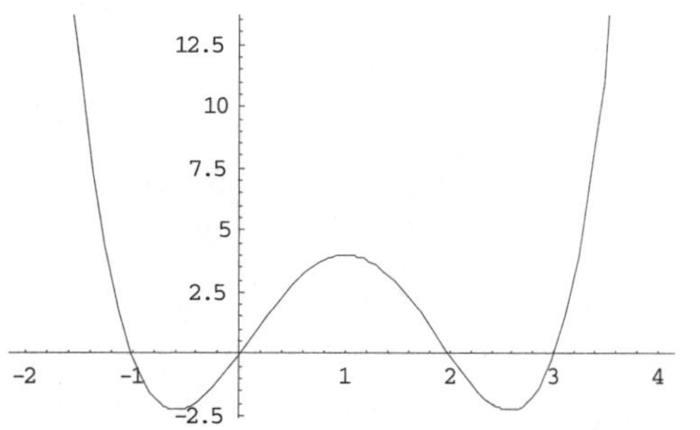

For Problems 9 – 14, use the equation of the polynomial function to:

a. state the degree and leading coefficient
b. predict the end behavior of the graph of the function
c. support your answer by graphing the function

9. $f(x) = 3x^3 + 2x$

10. $f(x) = -0.01x^4 + 0.02x^2 + 0.15$

11. $f(x) = (3x-1)^2(2-x)$

12. $f(x) = (2x+1)(4x-3)(5x+4)$

13. $f(x) = (3-4x)(5+4x)(1-x)$

14. $f(x) = -\frac{1}{2}(3-4x)(5+4x)(1-x)^2$

15. a. Graph the function $f(x) = x^3 - 27x$ using the window [-10, 10] by [-10. 10].
    b. Graph the function again so the graph is complete, showing the correct number of turning points.

16. a. Graph the function $f(x) = 27x^3 - x$ using the window [-10, 10] by [-10. 10].
    b. Graph the function again so the graph is complete, showing the correct number of turning points.

17. a. Graph the function $f(x) = 5x^4 - 10x^2 - 80$ using the window [-10, 10] by [-10. 10].
    b. Graph the function again so the graph is complete, showing the correct number of turning points.

Section 6.2 Modeling Cubic and Quartic Functions

1. Find the cubic function that models the data in the table below.

x	-2	-1	0	1	2	3	4
y	-25	-12	-5	2	15	40	83

2. Find the cubic function that models the data in the table below.

x	-2	-1	0	1	2	3	4
y	120	30	0	-30	-120	-330	-720

3. Find the quartic function that models the data in the table below.

x	-2	-1	0	1	2	3	4
y	32	-13	-20	-13	32	187	572

4. Find the quartic function that models the data in the table below.

x	-2	-1	0	1	2	3	4
y	1.8	.15	0	.15	1.8	8.55	26.4

5. a. Make a scatter-plot of the data in the table below.

x	-3	-2	-1	0	1	2	3	4	5	6
y	-11.9	-2.4	1.1	1.6	2.1	5.6	15.2	34.1	63.8	108.5

   b. Does it appear that a linear model or a cubic model is a better fit for the data?

6. a. Find a cubic function that models the data in Problem 5.
   b. Find a linear function that models the data in Problem 5.
   c. Visually determine which function is the better fit for the data.

7. a. Make a scatter-plot of the data in the table below.

x	-3	-2	-1	0	1	2	3
y	15	-10	-2	5	-3	-11	14

   b. Does it appear that a cubic model or a quartic model is a better fit for the data?

8. a. Find a cubic function that models the data in Problem 7.
   b. Find a quartic function that models the data in Problem 7.
   c. Visually determine which function is the better fit for the data.

9. a. Make a scatter-plot of the data in the table below.

x	-12	-10	-8	-5	-3	0	2	5	10	15	17
y	-35	-25	-15.05	-5.1	0.58	3.68	5.17	6.12	5.77	6.06	6.93

   b. Does it appear that a linear model or a cubic model is a better fit for the data?

10. a. Find a cubic function that models the data in Problem 9.
    b. Find a linear function that models the data in Problem 9.
    c. Visually determine which function is the better fit for the data.

11. a. Make a scatter-plot of the data in the table below.

x	0	1	2	3	4	5	6	7	8
y	-23	-19	-18	-16	-12	-11	-6	10	40

    b. Does it appear that a cubic model or a quartic model is a better fit for the data?

12. a. Find a cubic function that models the data in Problem 11.
    b. Find a quartic function that models the data in Problem 11.
    c. Visually determine which function is the better fit for the data.

Section 6.3 Solution of Polynomial Equations

In Problems 1 – 9, solve the polynomial equations by factoring and check the solutions graphically.

1. $x^3 - 4x = 0$
2. $x^3 - 16x = 0$
3. $3x^3 - 18x^2 = 0$
4. $x^3 - x^2 - 30x = 0$
5. $x^3 + 2x^2 - 5x - 6 = 0$
6. $x^4 - 8x^3 = 0$
7. $x^4 + 2x^3 + x^2 = 0$
8. $x^4 - 8x^3 + 15x^2 = 0$
9. $x^4 + 7x^3 + 6x^2 = 0$
10. $x^4 + x^3 - 8x^2 - 12x = 0$

In Problems 11 – 14, use factoring by grouping to solve the equations.

11. $x^3 + 3x^2 - 4x - 12 = 0$
12. $x^3 - 9x + 2x^2 - 18 = 0$
13. $x^3 - 7x^2 - x + 7 = 0$
14. $2x^3 - 50x + x^2 - 25 = 0$

In Problems 15 – 18, use the root method to solve the equations.

15. $5x^3 - 40 = 0$
16. $\frac{1}{2}x^3 - 32 = 0$
17. $3x^4 - 243 = 0$
18. $\frac{1}{4}x^4 - 4 = 0$

In Problems 19 – 26, use factoring and the root method to solve the polynomial equations.

19. $2x^3 - \frac{1}{2}x = 0$
20. $6x^3 - 20x = 0$
21. $\frac{1}{3}x^3 - \frac{1}{2}x = 0$
22. $\frac{1}{2}x^3 - 28x = 0$
23. $0.2x^3 - 6x = 0$
24. $3x^4 - 24x^2 + 48 = 0$
25. $x^4 - 20x^2 + 100 = 0$
26. $x^4 - 0.5x^2 + 0.0625 = 0$

In Problems 27 and 28, find the solutions graphically.

27. $x^3 - 7x^2 + 6x + 9 = 0$
28. $\frac{1}{2}x^3 - 4x + 1 = 0$

Section 6.4 Polynomial Equations Continued; Fundamental Theorem of Algebra

In Problems 1 – 4, use synthetic division to find the quotient and remainder.

1. $(x^4 - 3x^3 + 2x^2 + 4x + 18) \div (x+2)$

2. $(3x^4 - 2x^3 - 60x - 5) \div (x-3)$

3. $(x^5 + 3x^3 + 6x^2 - 5) \div (x+1)$

4. $(x^5 + 3x^2 + 2) \div (x+2)$

In Problems 5 and 6, determine whether the given constant is a solution to the given polynomial equation.

5. $3x^4 - 6x^3 + 2x^2 + x - 10 = 0;\ 2$

6. $2x^4 - 5x^2 - 3x - 120 = 0;\ -3$

In Problems 7 and 8, determine whether the second polynomial is a factor of the first polynomial.

7. $x^4 + 3x^3 - 5x^2 + 6x - 7;\ x+1$

8. $-x^4 + 4x^3 - 6x^2 + 3x + 2;\ x-2$

In Problems 9 – 12, one or more solutions of a polynomial equation are given. Use synthetic division to find any remaining solutions.

9. $x^3 - 3x^2 - 13x + 15 = 0;\ 1$

10. $x^3 + 8x^2 + 19x + 12 = 0;\ -1$

11. $x^4 - 2x^3 - 28x^2 + 8x + 96 = 0;\ 2, -2$

12. $x^4 - 3x^3 - 11x^2 + 3x + 10 = 0;\ 5, 1$

In Problems 13 – 16, find one solution graphically, then find the remaining solutions using synthetic division.

13. $2x^3 + 9x^2 + 7x - 6 = 0$

14. $x^3 + 2x^2 - 3x = 0$

15. $4x^3 + x^2 - 64x - 16 = 0$

16. $x^3 - 11x^2 + 16x + 84 = 0$

In Problems 17 and 18, find all possible rational solutions of the polynomial equations.

17. $x^3 + 2x^2 - 4x + 12 = 0$

18. $4x^3 + 6x^2 - 5x + 10 = 0$

In Problems 19 – 22, find all rational zeros.

19. $x^3 - x^2 - 14x + 24 = 0$

20. $x^3 + 3x^2 - 4x - 12 = 0$

21. $x^3 - x^2 - 5x - 3 = 0$

22. $x^4 - x^3 - 12x^2 + 28x - 16 = 0$

In Problems 23 – 28, solve each of the equations exactly in the complex number system. (Find one solution graphically and then use the quadratic formula to find remaining solutions.)

23. $x^3 - 5x^2 + 4x - 20 = 0$

24. $x^3 + x^2 + 5x + 5 = 0$

25. $x^3 + 2x^2 - x + 6 = 0$

26. $x^3 - 4x^2 + 9x - 36 = 0$

27. $x^3 + 5x^2 + 9x + 6 = 0$

28. $2x^3 + x^2 + 2x - 12 = 0$

Section 6.5 Rational Functions and Rational Equations

In Problems 1 – 6, give any (a) vertical and (b) horizontal asymptotes for the graphs of the rational functions.

1. $f(x)=\dfrac{6}{x+2}$

2. $f(x)=\dfrac{5}{x+1}$

3. $f(x)=\dfrac{x+3}{2x+5}$

4. $f(x)=\dfrac{3+2x}{4-x}$

5. $f(x)=\dfrac{x^3-2}{x^2+4}$

6. $f(x)=\dfrac{x^2+1}{x^2+5}$

7. Which of the following have no vertical asymptote?

a. $f(x)=\dfrac{x+2}{x^2-1}$

b. $f(x)=\dfrac{4}{x^2+2x}$

c. $f(x)=\dfrac{4x^2+1}{x^2+4}$

d. $f(x)=\dfrac{x+2}{(x-3)(x+1)}$

8. Which of the following functions have no horizontal asymptote?

a. $f(x)=\dfrac{3x+1}{x-2}$

b. $f(x)=\dfrac{2x}{x^2-4}$

c. $f(x)=\dfrac{x^3}{2-x^2}$

d. $f(x)=\dfrac{4x}{(x+2)(x-1)}$

In Problems 9 – 12, find (a) the vertical asymptote, (b) the horizontal asymptote, and (c) sketch the graph of the function.

9. $f(x)=\dfrac{x-4}{x+2}$

10. $f(x)=\dfrac{3x}{x-6}$

11. $f(x)=\dfrac{2x+1}{9-x^2}$

12. $f(x)=\dfrac{x^2}{1-x}$

13. a. Graph $f(x)=\dfrac{x+3}{x^2-6x+8}$

    b. Find y when $x=-2$ and $x=0$.

    c. Find x when $y=0$.

    d. Solve $-6=\dfrac{x+3}{x^2-6x+8}$ using analytical methods.

14. a. Graph $f(x)=\dfrac{x+1}{x^2-9}$

    b. Find y when $x=-2$ and $x=2$.

    c. Find x when $y=-0.25$.

    d. Solve $-\dfrac{3}{5}=\dfrac{x+1}{x^2-9}$ using analytical methods.

15. a. Graph $f(x)=\dfrac{x+3}{5-x}$

    b. Find y when $x=-2$ and $x=1$.

    c. Find x when $y=7$.

    d. Solve $3=\dfrac{x+3}{5-x}$ using analytical methods.

16. Use analytical methods to solve $x-\dfrac{x^2+2}{x}=\dfrac{1}{x}+2$.

17. Use analytical methods to solve $1+\dfrac{2}{x-1}=\dfrac{3x}{1-x}$.

Section 6.6 Polynomial and Rational Inequalities

In Problems 1 – 12, use analytical and graphical methods to solve the inequalities.

1. $3x^4 - x^3 \geq 0$

2. $4x^4 - x^2 \leq 0$

3. $x^3 - 9x \leq 0$

4. $(x-1)(x+3) > 0$

5. $(x+2)^2(x-4) \leq 0$

6. $(2x-1)(x-3)^3 > 0$

7. $\dfrac{x}{x+4} < 0$

8. $\dfrac{2x+4}{x} > 6$

9. $(x-2)^3 > 8$

10. $(x+4)^3 < 64$

11. $\dfrac{4}{x} + \dfrac{x-2}{4} \geq 2$

12. $\dfrac{x-1}{x} - \dfrac{x}{x+1} < \dfrac{1}{x}$

For Problems 13 – 15, use the graph of $y = f(x)$ to solve the requested inequality.

13.

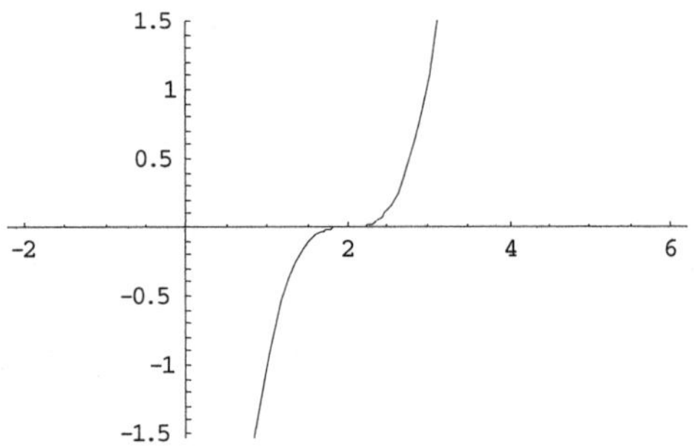

a. $f(x) < 0$

b. $f(x) \geq 0$

14.

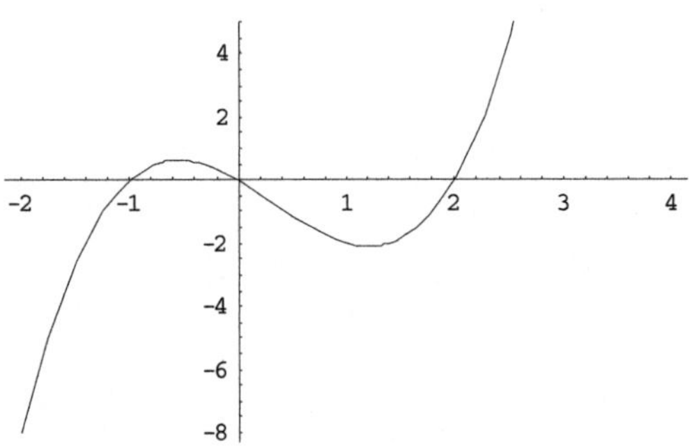

a. $f(x) < 0$

b. $f(x) \geq 0$

15.

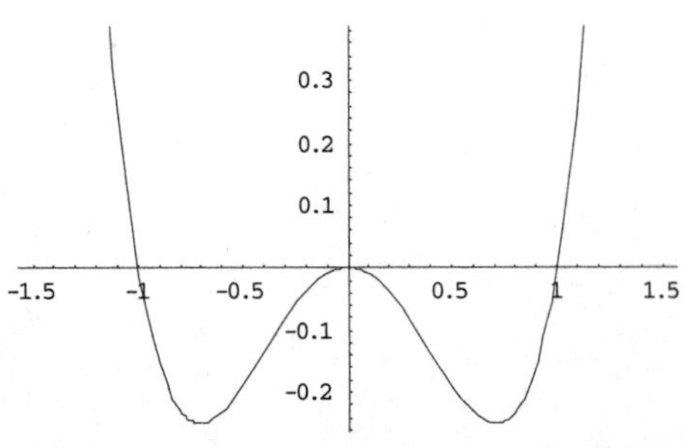

a. $f(x) < 0$

b. $f(x) \geq 0$

Section 7.1 Systems of Linear Equations in Three Variables

In Problems 1 – 4, solve the systems of equations.

1. $\begin{cases} x+2y-z=10 \\ y+3z=18 \\ z=4 \end{cases}$

2. $\begin{cases} x-2y-2z=-3 \\ y+4z=-8 \\ z=-3 \end{cases}$

3. $\begin{cases} x-y+5z=12 \\ y+2z=2 \\ z=1 \end{cases}$

4. $\begin{cases} x+3y-z=-1 \\ y+4z=3 \\ z=1 \end{cases}$

In Problems 5 – 14, use left-to-right elimination to solve the systems of equations.

5. $\begin{cases} x-y-2z=-3 \\ y+z=1 \\ 2y-z=-10 \end{cases}$

6. $\begin{cases} x-y-z=12 \\ y+2z=4 \\ y-z=10 \end{cases}$

7. $\begin{cases} 2x+4y+z=19 \\ y-z=4 \\ y+z=2 \end{cases}$

8. $\begin{cases} x+2y+4z=6 \\ y+2z=0 \\ y-z=-3 \end{cases}$

9. $\begin{cases} x+y-z=-2 \\ x-y+z=12 \\ x-y-z=0 \end{cases}$

10. $\begin{cases} x+2y+z=1 \\ 2x-y+z=0 \\ -x+y-z=-1 \end{cases}$

11. $\begin{cases} 2x-y+z=8 \\ x-4y+2z=21 \\ 3x-2y-z=9 \end{cases}$

12. $\begin{cases} -x+4y+2z=7 \\ 2x-2y+z=-1 \\ 3x+2y+z=0 \end{cases}$

13. $\begin{cases} 4x+y-z=0 \\ 2x-y-2z=6 \\ 6x-2y-2z=-1 \end{cases}$

14. $\begin{cases} x+y+z=3 \\ x-y+z=11 \\ x+2y-z=-5 \end{cases}$

The systems in Problems 15 – 20 have nonunique solutions. Solve each system, if possible.

15. $\begin{cases} x+2y+z=6 \\ y-z=4 \end{cases}$

16. $\begin{cases} x+2y-z=3 \\ y+3z=2 \end{cases}$

17. $\begin{cases} x+y+4z=4 \\ 2x+y+5z=6 \\ -x+y+2z=0 \end{cases}$

18. $\begin{cases} x-3y-2z=-8 \\ \quad y+4z=10 \\ \qquad 0=6 \end{cases}$

19. $\begin{cases} x+2y+3z=4 \\ 3x+6y+z=2 \\ 2x+4y+z=-2 \end{cases}$

20. $\begin{cases} x-2y+3z=4 \\ 3x-y-z=2 \\ x+y-3z=-2 \end{cases}$

Section 7.2 Matrix Solution of Systems of Linear Equations

In Problems 1 – 4, write the augmented matrix associated with the given system.

1. $\begin{cases} x+2y-z=4 \\ 2x+y+4z=8 \\ -x+3y-2z=6 \end{cases}$

2. $\begin{cases} 3x+2y+z=8 \\ -2x+3y+4z=2 \\ x+4y-z=6 \end{cases}$

3. $\begin{cases} 2x+4y+5z=11 \\ 3x-y+2z=4 \\ 5x+y-z=5 \end{cases}$

4. $\begin{cases} 3x+2y+2z=7 \\ 5-2y+4z=7 \\ 8-2y-z=5 \end{cases}$

In Problems 5 – 14, the matrix associated with the solution to a system of linear equations in *x*, *y*, *z* is given. Write the solution to the system, if it exists.

5. $\left[\begin{array}{ccc|c} 1 & 0 & 0 & 3 \\ 0 & 1 & 0 & -2 \\ 0 & 0 & 1 & 4 \end{array}\right]$

6. $\left[\begin{array}{ccc|c} 1 & 0 & 0 & -2 \\ 0 & 1 & 0 & 6 \\ 0 & 0 & 1 & 9 \end{array}\right]$

7. $\left[\begin{array}{ccc|c} 1 & 0 & 0 & -3 \\ 0 & 2 & 0 & 4 \\ 0 & 0 & 4 & -4 \end{array}\right]$

8. $\left[\begin{array}{ccc|c} -1 & 1 & 0 & 2 \\ 1 & 0 & 1 & -4 \\ 3 & 1 & 0 & -2 \end{array}\right]$

9. $\left[\begin{array}{ccc|c} 1 & -1 & 1 & 4 \\ 0 & 1 & 2 & 3 \\ -1 & 0 & 3 & -1 \end{array}\right]$

10. $\left[\begin{array}{ccc|c} 1 & 0 & 2 & 3 \\ 0 & 1 & 4 & 2 \\ 0 & 0 & 0 & 2 \end{array}\right]$

11. $\left[\begin{array}{ccc|c} 1 & 0 & 3 & 2 \\ 0 & 1 & 6 & 5 \\ 0 & 0 & 0 & 0 \end{array}\right]$

12. $\left[\begin{array}{ccc|c} 1 & 1 & -1 & 1 \\ 2 & 1 & 1 & 2 \\ 1 & 0 & 1 & 3 \end{array}\right]$

13. $\left[\begin{array}{ccc|c} 1 & 2 & 3 & 4 \\ 3 & 6 & 1 & 2 \\ 2 & 4 & 1 & -2 \end{array}\right]$

14. $\left[\begin{array}{ccc|c} 1 & 1 & 4 & -2 \\ 2 & 4 & -1 & 5 \\ 6 & -1 & 3 & 9 \end{array}\right]$

Solve the systems in Problems 15 – 18.

15. $\begin{cases} 2x \quad\quad +z=5 \\ \quad\quad 2y-z=-10 \\ 2x-y+z=8 \end{cases}$

16. $\begin{cases} 6x-3y+4z=-22 \\ 4x-2y+7z=-6 \\ 2x+5y+6z=26 \end{cases}$

17. $\begin{cases} x+y+z+w=7 \\ -x+2y-z+w=4 \\ x-y-z+2w=0 \\ -x+y+z-w=1 \end{cases}$

18. $\begin{cases} x+y+z+w=2 \\ -x+y-z-w=-2 \\ x-y+z+2w=1 \\ -x-y-z+w=-4 \end{cases}$

In Problems 19 – 22, find the solutions, if they exist, to the systems.

19. $\begin{cases} 2x \quad\quad +z=0 \\ -x+y+z=-1 \\ 3x+2y \quad\quad =-11 \end{cases}$

20. $\begin{cases} 2x+y+7z=9 \\ x+y+5z=5 \\ -x+y+z=-3 \end{cases}$

21. $\begin{cases} x+y+6z=2 \\ 2x+y+7z=5 \\ -x+y+4z=-4 \end{cases}$

22. $\begin{cases} 2x-3y+4z=4 \\ x+3y-2z=0 \\ x+6y+2z=-3 \end{cases}$

Section 7.3 Matrix Operations

1. Find the missing variables $\begin{bmatrix} x & 2 & 2y \\ z+w & 3 & 3x \\ -1 & z-w & 6 \end{bmatrix} = \begin{bmatrix} 3 & 1 & 4 \\ 2 & 3 & 9 \\ -1 & 4 & 6 \end{bmatrix}$.

For Problems 2 – 11, use the following matrices and then perform the indicated operation, if possible.

$$A = \begin{bmatrix} 1 & 3 & -1 \\ 2 & 0 & 4 \\ -2 & 5 & 2 \end{bmatrix} \quad B = \begin{bmatrix} -2 & 3 & 4 \\ 1 & -6 & 2 \end{bmatrix} \quad C = \begin{bmatrix} 1 & 2 \\ -3 & 4 \\ 0 & 3 \end{bmatrix} \quad D = \begin{bmatrix} 4 & 2 \\ -1 & 3 \end{bmatrix} \quad E = \begin{bmatrix} 3 & 0 \\ 2 & 4 \end{bmatrix}$$

2. $2D+E$
3. $E-3D$
4. $2C+B$
5. AC
6. BE
7. EB
8. $3BC$
9. DC
10. CDB
11. E^2D

For Problems 12 – 15, use $A = \begin{bmatrix} 5 & 0 \\ -2 & 1 \end{bmatrix}$ and $B = \begin{bmatrix} 1 & 4 \\ 2 & 3 \end{bmatrix}$.

12. Find A^2.
13. Find B^2
14. Find AB.
15. Find BA.

For Problems 16 – 19, use $A = \begin{bmatrix} 1 & 2 & 3 \\ 0 & 1 & 5 \\ -1 & -3 & 1 \end{bmatrix}$ and $B = \begin{bmatrix} 2 & 1 & 4 \\ 1 & 2 & -1 \\ 3 & -5 & 1 \end{bmatrix}$.

16. Find $2A+3B$.
17. Find A^2.
18. AB.
19. Find BA.

20. Two brands of coffee; Morning Glory and Wake Up Call, are made by blending three types of coffee beans (A, B, and C). The number of pounds of each type of bean in a 20 lb. blend of each brand of coffee are given below:

$$\begin{array}{r} \\ \textit{Morning Glory} \\ \textit{Wake Up Call} \end{array} \begin{array}{c} \begin{array}{ccc} \text{A} & \text{B} & \text{C} \end{array} \\ \begin{bmatrix} 12 & 4 & 4 \\ 8 & 5 & 7 \end{bmatrix} \end{array}$$

Use matrix multiplication and find the total cost of a 20 lb. bag of each blend of coffee if the cost per pound of beans A, B, and C are \$2.20, \$1.50, and \$0.80, respectively.

Section 7.4 Inverse Matrices; Matrix Equations

1. Suppose $A = \begin{bmatrix} 1 & 2 \\ 4 & 6 \end{bmatrix}$ and $B = \begin{bmatrix} -3 & 1 \\ 2 & -\frac{1}{2} \end{bmatrix}$.

a. Compute AB and BA, if possible.
b. Are A and B inverses of each other?

2. Suppose $A = \begin{bmatrix} 2 & 3 \\ 1 & 8 \end{bmatrix}$ and $B = \begin{bmatrix} -1 & 3 \\ 1 & -2 \end{bmatrix}$.

a. Compute AB and BA, if possible.
b. Are A and B inverses of each other?

For Problems 3 – 10, find the inverse of the given matrix.

3. $\begin{bmatrix} 2 & 4 \\ 1 & -1 \end{bmatrix}$

4. $\begin{bmatrix} 2 & 4 \\ -3 & 1 \end{bmatrix}$

5. $\begin{bmatrix} -1 & -6 \\ 3 & 5 \end{bmatrix}$

6. $\begin{bmatrix} 2 & 6 \\ 1 & 3 \end{bmatrix}$

7. $\begin{bmatrix} 8 & 2 \\ 7 & 2 \end{bmatrix}$

8. $\begin{bmatrix} 1 & 0 & 1 \\ 2 & -2 & -1 \\ 3 & 0 & 0 \end{bmatrix}$

9. $\begin{bmatrix} 1 & 4 & 3 \\ 1 & -3 & -2 \\ 2 & 5 & 4 \end{bmatrix}$

10. $\begin{bmatrix} 2 & 4 & 6 \\ -1 & -4 & -3 \\ 0 & 1 & -1 \end{bmatrix}$

11. If $A^{-1} = \begin{bmatrix} 1 & 2 \\ 4 & 6 \end{bmatrix}$, solve $AX = \begin{bmatrix} 3 \\ -1 \end{bmatrix}$ for X.

12. If $A^{-1} = \begin{bmatrix} 1 & 2 & 0 \\ -1 & 3 & 1 \\ 2 & 4 & 1 \end{bmatrix}$, solve $AX = \begin{bmatrix} -2 \\ 1 \\ 4 \end{bmatrix}$ for X.

13. Solve the matrix equation $\begin{bmatrix} 1 & -1 \\ 5 & -6 \end{bmatrix} \begin{bmatrix} x \\ y \end{bmatrix} = \begin{bmatrix} 2 \\ 4 \end{bmatrix}$.

14. Solve the matrix equation $\begin{bmatrix} -1 & -2 & -3 \\ -1 & 4 & 6 \\ 1 & -1 & -2 \end{bmatrix} \begin{bmatrix} x \\ y \\ z \end{bmatrix} = \begin{bmatrix} -20 \\ 7 \\ 4 \end{bmatrix}$.

In Problems 15 – 19, use an inverse matrix to find the solution to the systems.

15. $\begin{cases} x+3y-6z=7 \\ 2x-y+2z=0 \\ x+y+2z=-1 \end{cases}$

16. $\begin{cases} x+2y+z=5 \\ 2x+y-3z=-2 \\ 3x+y+4z=-5 \end{cases}$

17. $\begin{cases} x+2y+3z=5 \\ 2x+3y+2z=2 \\ -x-2y-4z=-1 \end{cases}$

18. $\begin{cases} 2x+y+3z-2w=-6 \\ 4x+3y+z-w=-2 \\ x+y+z+w=-5 \\ -2x-2y+2z+2w=-10 \end{cases}$

19. $\begin{cases} x+y+z+w=-2 \\ -x+4y+z-w=7 \\ x-2y+z-2w=-2 \\ -x-2y+z+2w=-2 \end{cases}$

Section 7.5 Systems of Non-linear Equations

In Problems 1 – 6, solve the system algebraically.

1. $\begin{cases} x^2 - y = 2 \\ 4x - y = 5 \end{cases}$

2. $\begin{cases} x^2 - 2y = -7 \\ 3x - 2y = -5 \end{cases}$

3. $\begin{cases} x^2 + 5y = 16 \\ x + y = 2 \end{cases}$

4. $\begin{cases} x^2 + y^2 = 5 \\ y = 2x \end{cases}$

5. $\begin{cases} x + y = 7 \\ xy = 10 \end{cases}$

6. $\begin{cases} 2xy + y = 7 \\ 2x - y = 5 \end{cases}$

In Problems 7 – 12, solve the system algebraically or graphically.

7. $\begin{cases} x^2 + 4x - y = 43 \\ 2x - y + 6 = 14 \end{cases}$

8. $\begin{cases} x^2 - 2x - y = -2 \\ 4x - y + 2 = 8 \end{cases}$

9. $\begin{cases} x^2 - 2x + y = 9 \\ 3x - y - 4 = 7 \end{cases}$

10. $\begin{cases} x^2 + 4x + 2y = 3 \\ 2x - 2y + 1 = -7 \end{cases}$

11. $\begin{cases} x^2 + 6y = 28 \\ \sqrt{y} - x = 0 \end{cases}$

12. $\begin{cases} xy + y = 15 \\ 2x + y = 11 \end{cases}$

In Problems 13 – 15, solve the system graphically.

13. $\begin{cases} y = x^3 - 5x + 3 \\ y = -x + 3 \end{cases}$

14. $\begin{cases} y = x^3 - 6x^2 + 3x \\ y = x^2 + 4x - 7 \end{cases}$

15. $\begin{cases} x^3 + x^2 + y = 8 \\ 2x^2 + y = 8 \end{cases}$

Section 8.1 Systems of Inequalities

For Problems 1 – 6, graph each inequality.

1. $y \le 2x+3$
2. $y > -x+6$
3. $2x+4y \le 8$
4. $4x-2y \le 12$
5. $\frac{x}{2}+\frac{x}{4} \ge 2$
6. $5x+3y \ge 15$

For Problems 7 and 8, use the following constraints.

$$\begin{cases} x+y \le 4 \\ 3x+y \le 6 \\ x \ge 0,\ y \ge 0 \end{cases}$$

7. Graph the solution region.

8. Find the corners of the solution region.

For Problems 9 and 10, use the following constraints.

$$\begin{cases} 2x+y \le 10 \\ x+2y \le 8 \\ x \ge 0,\ y \ge 0 \end{cases}$$

9. Graph the solution region.

10. Find the corners of the solution region.

For Problems 11 and 12, use the following constraints.

$$\begin{cases} 5x+4y \le 20 \\ x+2y \le 7 \\ x \ge 0,\ y \ge 0 \end{cases}$$

11. Graph the solution region.

12. Find the corners of the solution region.

For Problems 13 and 14, use the following constraints.

$$\begin{cases} 4x+3y\le 24 \\ 2x+y\le 10 \\ x\ge 0,\ y\ge 0 \end{cases}$$

13. Graph the solution region.

14. Find the corners of the solution region.

For Problems 15 and 16, use the following constraints.

$$\begin{cases} x+y\le 7 \\ x+y\ge 2 \\ -3x+2y\le 4 \\ -3x+2y\ge -9 \\ x\ge 0,\ y\ge 0 \end{cases}$$

15. Graph the solution region.

16. Find the corners of the solution region.

For Problems 17 and 18, use the following constraints.

$$\begin{cases} 5x+3y\le 30 \\ -x+3y\le 12 \\ x+y\ge 4 \\ x\ge 0,\ y\ge 0 \end{cases}$$

17. Graph the solution region.

18. Find the corners of the solution region.

For Problems 19 and 20, use the following constraints.

$$\begin{cases} 2x+3y\le 12 \\ 4x+y\le 14 \\ 3x+y\ge 3 \\ x\ge 0,\ y\ge 0 \end{cases}$$

19. Graph the solution region.

20. Find the corners of the solution region.

In Problems 21- 24, graph the solution of the systems of inequalities.

21. $\begin{cases} x^2 - y < 6 \\ x - y > 4 \end{cases}$

22. $\begin{cases} xy \geq 8 \\ x^2 - y - 6x < -9 \end{cases}$

23. $\begin{cases} x^2 - y < 0 \\ 2x + y > 6 \end{cases}$

24. $\begin{cases} xy \leq 6 \\ x + y \leq 5 \end{cases}$

Section 8.2 Linear Programming: Graphical Methods

For Problems 1 – 4, use the following constraints.

$$\begin{cases} x + y \le 8 \\ 2x + y \le 13 \\ x \ge 0,\ y \ge 0 \end{cases}$$

1. Graph the region that corresponds to the above system.

2. Find the coordinates of all the corner points for the region.

3. Maximize (if possible) the quantity $z = 4x + 3y$ and give the ordered pair that gives the value.

4. Minimize (if possible) the quantity $q = 4x + 3y$ and give the ordered pair that gives the value.

For Problems 5 – 8, use the following constraints.

$$\begin{cases} 2x + y \le 5 \\ -x + 2y \le 4 \\ 4x + y \ge 1 \\ x \ge 0,\ y \ge 0 \end{cases}$$

5. Graph the region that corresponds to the above system.

6. Find the coordinates of all the corner points for the region.

7. Maximize (if possible) the quantity $z = x + y$ and give the ordered pair that gives the value.

8. Minimize (if possible) the quantity $q = x + y$ and give the ordered pair that gives the value.

For Problems 9 – 12, use the following constraints.

$$\begin{cases} x + y \le 4 \\ x + 6y \ge 6 \\ 2x + y \le 5 \\ x \ge 0,\ y \ge 0 \end{cases}$$

9. Graph the region that corresponds to the above system.

10. Find the coordinates of all the corner points for the region.

11. Maximize (if possible) the quantity $z = 4x + 5y$ and give the ordered pair that gives the value.

12. Minimize (if possible) the quantity $q = 4x + 5y$ and give the ordered pair that gives the value.

For Problems 13 – 18, use the following constraints.

$$\begin{cases} -4x + y \le -2 \\ 3x + y \le 6 \\ -x + y \ge 0 \\ x \ge 0,\ y \ge 0 \end{cases}$$

13. Graph the region that corresponds to the above system.

14. Find the coordinates of all the corner points for the region.

15. Maximize (if possible) the quantity $z = 5x + 4y$ and give the ordered pair that gives the value.

16. Minimize (if possible) the quantity $q = 2x + 6y$ and give the ordered pair that gives the value.

17. Maximize (if possible) the quantity $z = 4x$ and give the ordered pair that gives the value.

18. Minimize (if possible) the quantity $q = 8y$ and give the ordered pair that gives the value.

Section 8.3 Sequences and Discrete Functions

1. Find the first 5 terms of the sequence defined by $f(n) = 3n + 2$.

2. Find the first 5 terms of the sequence defined by $f(n) = (-1)^n (n+1)^2$.

3. Find the first 5 terms of the sequence defined by $f(n) = \frac{1}{n+1} + n$.

4. Find the first 5 terms of the sequence defined by $f(n) = \frac{60}{n}$.

5. Find the next 3 terms of the arithmetic sequence $4, 7, 10, 13, \ldots$

6. Find the next 3 terms of the arithmetic sequence $-2, -5, -8, -11, \ldots$

7. Find the 10th term of the arithmetic sequence with first term -2 and common difference 5.

8. Find the 30th term of the arithmetic sequence with first term 8 and common difference 3.5.

9. Find the next 3 terms of the geometric sequence $\frac{1}{2}, \frac{3}{2}, \frac{9}{2}, \frac{27}{2}, \ldots$

10. Find the next 3 terms of the geometric sequence $400, 300, 225, 168.75, \ldots$

11. Find the 10th term of the geometric sequence with first term 20 and common ratio 0.4.

12. Find the 40th term of the geometric sequence with first term 10 and common ratio 1.5.

13. Find the first 4 terms of the sequence with first term 6 and nth term $a_n = a_{n-1} + 4$.

14. Find the first 4 terms of the sequence with first term 10 and nth term $a_n = a_{n-1} - 2$.

15. Find the first 4 terms of the sequence with first term 8 and nth term $a_n = \frac{1}{2} a_{n-1} + 6$.

Section 8.4 Series

1. Find the sum of the first 8 terms of the sequence 2, 6, 10, 14, . . .

2. Find the sum of the first 10 terms of the sequence 16, 8, 4, 2, . . .

3. Find the sum of the first 20 terms of the arithmetic sequence -3, -7, -11, -15, . . .

4. Find the sum of the first 30 terms of the arithmetic sequence 4, 6.5, 9, 11.5, . . .

5. Find the sum of the first 25 terms of the arithmetic sequence with first term -3 and a common difference of 5.

6. Find the sum of the first 50 terms of the arithmetic sequence with first term 18 and a common difference of -3.

7. Find the sum of the first 20 terms of the geometric sequence 2, 6, 18, 54, . . .

8. Find the sum of the first 25 terms of the geometric sequence 16, 4, 1, ¼, . . .

9. Find the sum of the first 20 terms of the geometric sequence with first term 20 and ratio 3/2.

10. Find the sum of the first 30 terms of the geometric sequence with first term -4 and ratio -1/5.

11. Find the sum represented by the series 50 + 10 + 2 + 0.4 + . . . , if possible.

12. Find the sum represented by the series 3 + 6 + 12 + 24 + . . . , if possible.

In Problems 13 – 20, find the sum, if possible.

13. $\sum_{i=1}^{8} 3^i$

14. $\sum_{i=1}^{10} 7^i$

15. $\sum_{i=1}^{8} \left(\frac{1}{4}\right)^i$

16. $\sum_{i=1}^{5} \left(\frac{2}{3}\right)^i$

17. $\sum_{i=1}^{5} \left(-\frac{1}{6}\right)^i$

18. $\sum_{i=1}^{20} \left(\frac{3}{2}\right)^i$

19. $\sum_{i=1}^{\infty} \left(\frac{3}{5}\right)^i$

20. $\sum_{i=1}^{\infty} \left(\frac{3}{4}\right)^i$

Section 8.5 Preparing for Calculus

In Problems 1 – 4, the expressions represent the derivative of a function. Simplify the expressions.

1. $f'(x)=4(3x+5)^3(2x-9)^5+5(3x+5)^4(2x-9)^4$

2. $f'(x)=4(6x+6)(3x^2+6x-9)^3$

3. $f'(x)=5(x^2+6)^4(2x)(3x-8)^3+9(x^2+6)^5(3x-8)^2$

4. $f'(x)=\dfrac{2(x+3)-(x+3)^2}{(x-5)^2}$

In Problems 5 – 8, the derivative of $f(x)$ is denoted by $f'(x)$, and the value of the derivative at $x=a$ is $f'(a)$. Evaluate the derivatives at $x=a$.

5. Find $f'(3)$ if $f'(x)=6x^3-9x^2-6x+2$

6. Find $f'(-2)$ if $f'(x)=\dfrac{4}{x^2}-x$

7. Find $f'(1)$ if $f'(x)=\dfrac{2(x+4)(x-7)-(x+4)^2}{(x+7)^2}$

8. Find $f'(2)$ if $f'(x)=6e^{3x+4}$

In Problems 9 – 12, find each of the following
a. $f(x+h)$
b. $f(x+h)-f(x)$
c. $\dfrac{f(x+h)-f(x)}{h}$

9. $f(x)=6x-5$

10. $f(x)=4-2x$

11. $f(x)=\dfrac{x}{2}+4$

12. $f(x)=\dfrac{2-x}{3}$

Writing the equation defining a function in the form $y=f(x)$ usually makes it easier to find the derivative. Write each of the equations in Problems 13 – 16 in this form by solving for y.

13. $5x^3-8y=32$

14. $2y-9x^2=6$

15. $3x^2=4y-6x+2$

16. $3x^2+4y=8x+12$

17. The parabola $y=x^2-6x+2$ has associated equation $0=2x-6$.
    a. Find the solution of the associated equation to find the *x*-coordinate of the vertex of the parabola.
    b. Find the vertex of the parabola by using the answer from part (a).

18. The parabola $y=60x-6x^2$ has associated equation $0=60-12x$.
    a. Find the solution of the associated equation to find the *x*-coordinate of the vertex of the parabola.
    b. Find the vertex of the parabola by using the answer from part (a).

19. The parabola $y=3x^2+12x-5$ has associated equation $0=6x+12$.
    a. Find the solution of the associated equation to find the *x*-coordinate of the vertex of the parabola.
    b. Find the vertex of the parabola by using the answer from part (a).

20. Between $x=-5$ and $x=1$, the largest value of the expression $y=x^3+6x^2+2$ occurs where $3x^2+12x=0$.
    a. Find the values of *x* that satisfy $3x^2+12x=0$.
    b. Find the values of $y=x^3+6x^2+2$ at each of the solutions found in part (a).
    c. Which of the solutions from part (a) give the maximum value of $y=x^3+6x^2+2$ on [-5, 1]?

21. Between $x=-1.5$ and $x=3$, the largest value of the expression $y=3x-x^3$ occurs where $3-3x^2=0$.
    a. Find the values of *x* that satisfy $3-3x^2=0$.
    b. Find the values of $y=3x-x^3$ at each of the solutions found in part (a).
    c. Which of the solutions from part (a) give the maximum value of $y=3x-x^3$ on [-1.5, 3]?

To perform calculus operations on functions, it is frequently desirable or necessary to write the function in the form $y=cx^b$, where *b* is a constant. Write each of the functions in Problems 22 – 25 in this form.

22. $y=5\sqrt{x}$

23. $y=8\sqrt[5]{x}$

24. $y=7\sqrt[3]{x^4}$

25. $y=8\sqrt{x^5}$

In Problems 26 – 29, write each function without radicals using fractional exponents.

26. $y=\sqrt{4x+7}$

27. $y=\sqrt{(2x+3)^3}$

28. $y=\sqrt[3]{x^2-6}$

29. $y=\sqrt[5]{(4x+9)^2}$

In calculus it is often necessary to write expressions in the form cx^n, where n is a constant. In Problems 30 – 33, use negative exponents to write each term in the expressions in the form cx^n.

30. $y=\frac{1}{x^3}+\frac{1}{x^5}-\frac{6}{x^7}$

31. $y=\frac{4}{5x}+\frac{3}{x^3}-\frac{1}{2x^4}$

32. $y=\frac{2}{\sqrt{x}}+\frac{6}{\sqrt[3]{x}}$

33. $y=\frac{5}{\sqrt[3]{x^2}}$

Sometimes a derivative contains negative exponents, and simplification requires that all exponents be positive. In Problems 34 and 35, write each of the expressions without negative exponents.

34. $y=5x^{-2}-3x^{-6}$

35. $y=4x^{-1/2}$

CHAPTER 1 ANSWER KEY

Section 1.1

1. a. output b. input c. output
d. Domain = {-2, 1, 4, 7, 10, 13}
Range = {-8, 1, 10, 5, 3, 9}
e. For every value of *x*, there is only one value of *y*.

2. a. input b. output c. output
d. Domain = {-5, -1, 0, 6, 9, 15}
Range = {-2, 4, 1, 7, 2, -6}
e. For every value of *x* there is only one value of *y*.

3. a. 8 b. 15 c. 15

4. a. 0 b. -6 c. -2

5. $P(4)=1$, $P(3)=3$

6. $C(20)=1085$, $C(-10)=365$

7. a. 15/11 b. 2 c. 3/8

8. a. $\sqrt{9}=3$ b. $\sqrt{2}$ c. $\sqrt{1}=1$

9. yes 10. yes 11. yes

12. yes 13. yes 14. no

15. yes 16. no 17. yes

18. no 19. yes 20. no

21. no 22. yes 23. no

24. $A=\frac{1}{2}bh$

25. $W=4\left(\frac{x}{y}\right)+(D-C)$

26. Domain = {all real numbers}
Range = $\{y : y \geq -2\}$

27. Domain = $\{x : x \geq 0\}$
Range = $\{y : y \geq 0\}$

28. Domain = $\{x : x \neq -4\}$ and
Range = $\{y : y \neq 0\}$

29. a. Domain = $\{x : x \geq 2\}$
Range = $\{y : y \geq 0\}$
b. Domain = {all real numbers}
Range = {all real numbers}

30. a. Domain = $\{x : x \neq 4\}$
b. Domain = {all real numbers.

Section 1.2

1.

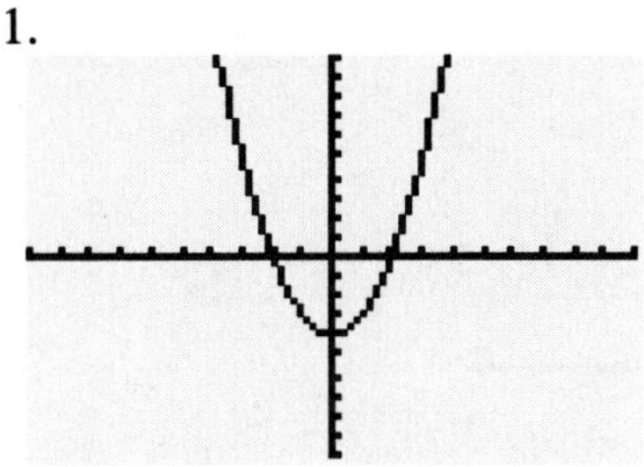

2.

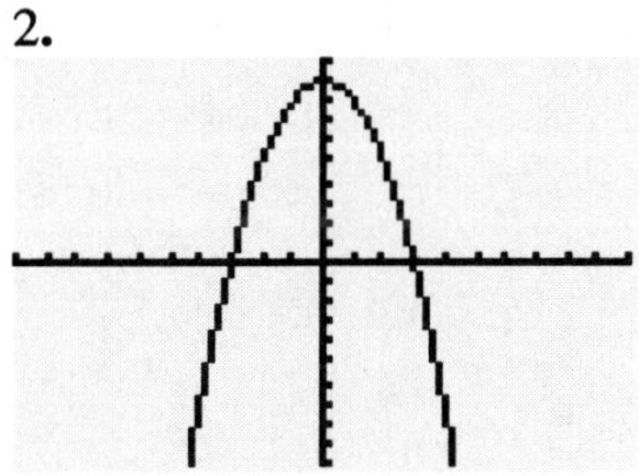

3.

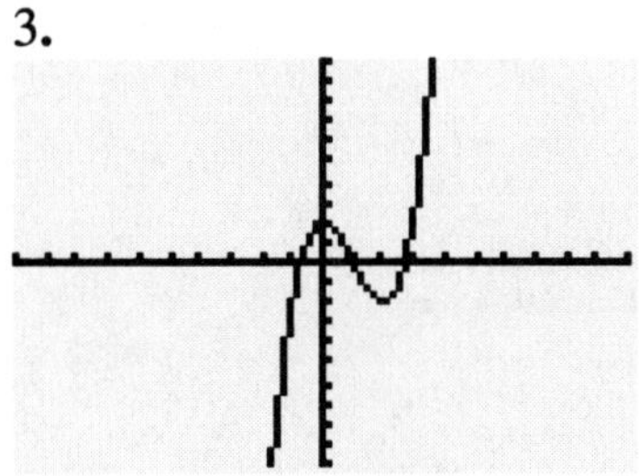

4.

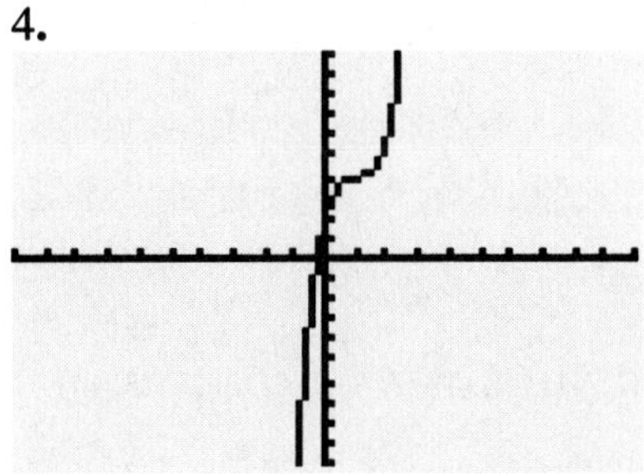

5.

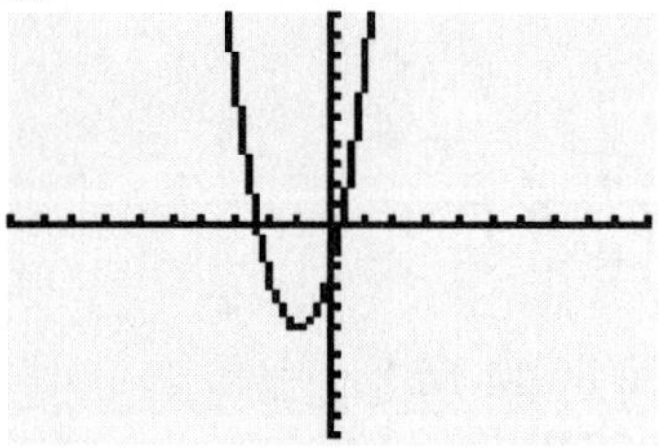

6.

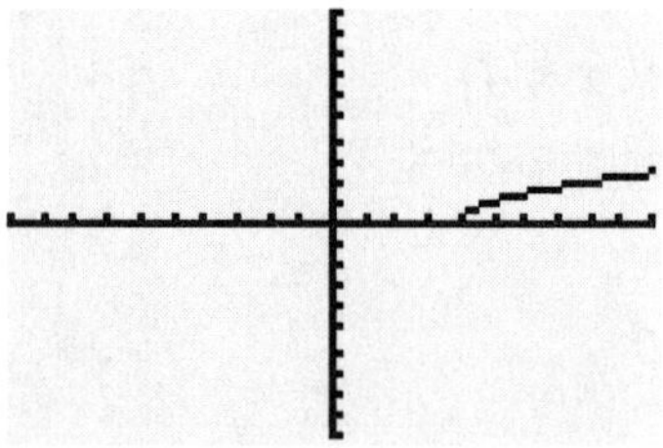

7.

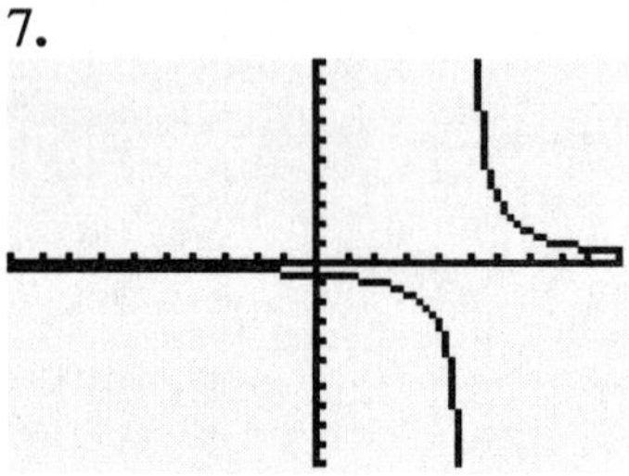

8.

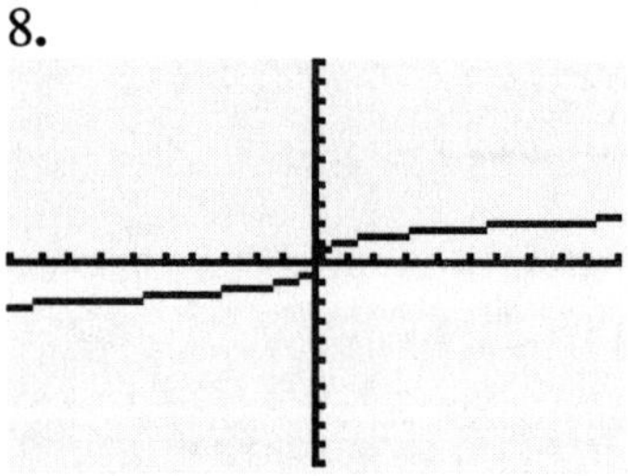

9.

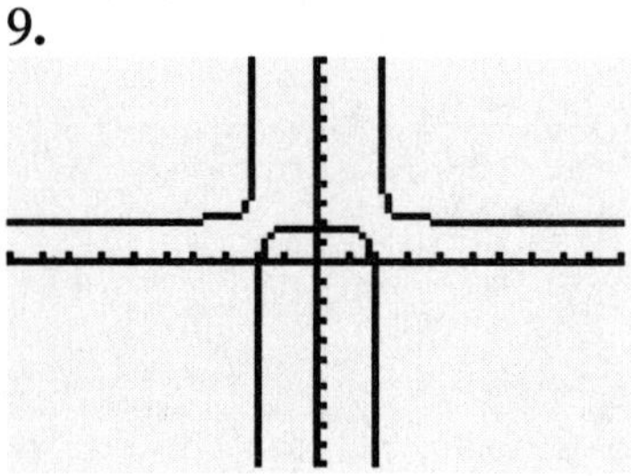

10.

x	-3	-2	-1	0	1	2	3
$y=3(x+1)^2-4$	8	-1	-4	-1	8	23	44
(x, y)	(-3, 8)	(-2, -1)	(-1, -4)	(0, -1)	(1, 8)	(2, 23)	(3, 44)

11.

x	-3	-2	-1	0	1	2	3
$y=\frac{4}{x+1}$	-2	-4	---	4	2	4/3	1
(x, y)	(-3,-2)	(-2, -4)	---	(0, 4)	(1, 2)	(2, 4/3)	(3, 1)

12. a.

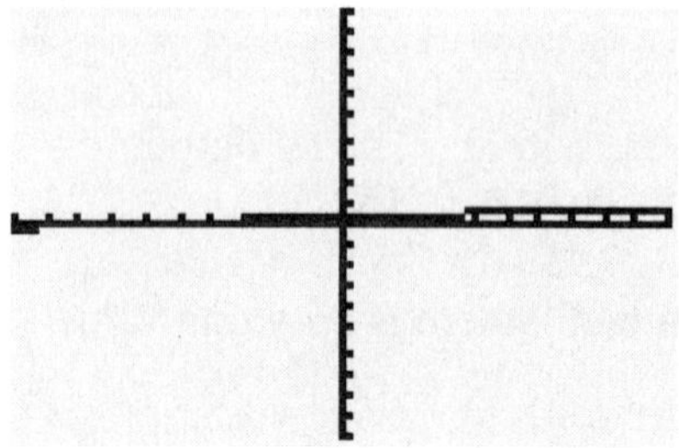

b.

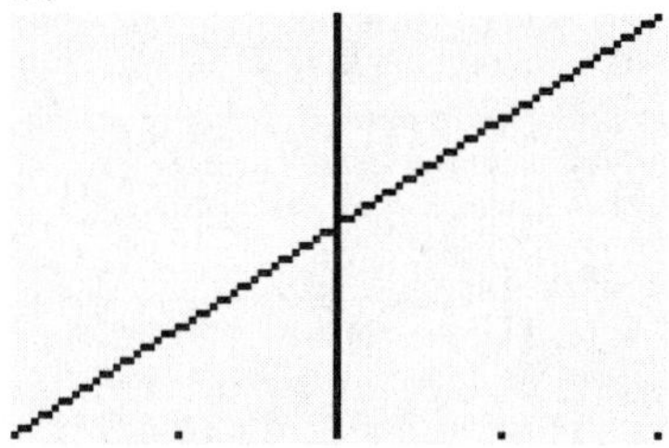

c. graph in part (b).

13. a.

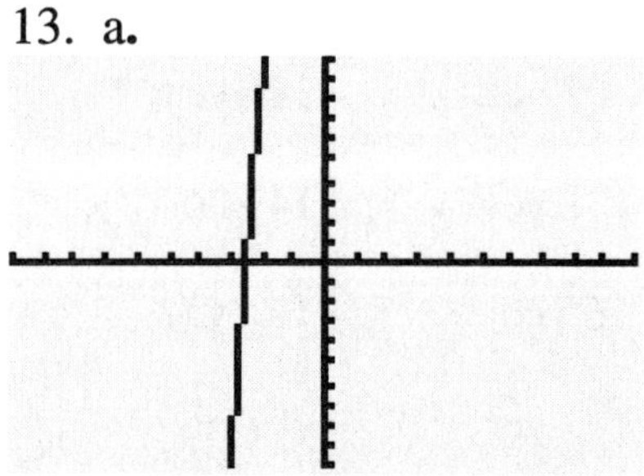

b.

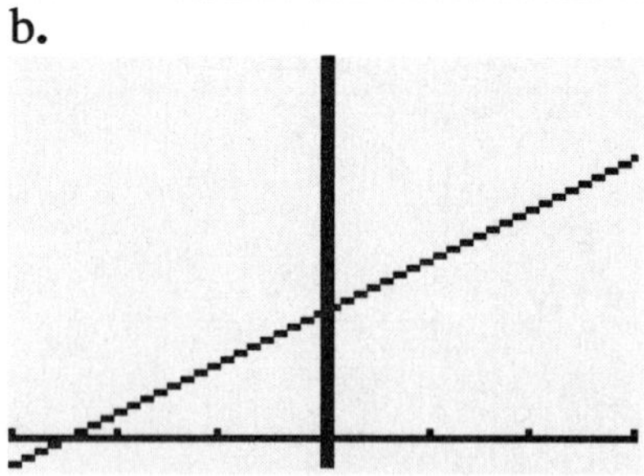

c. graph in part (b).

14. a.

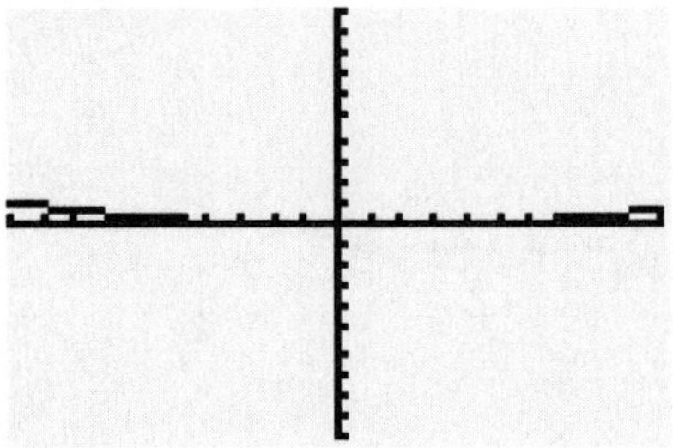

b.

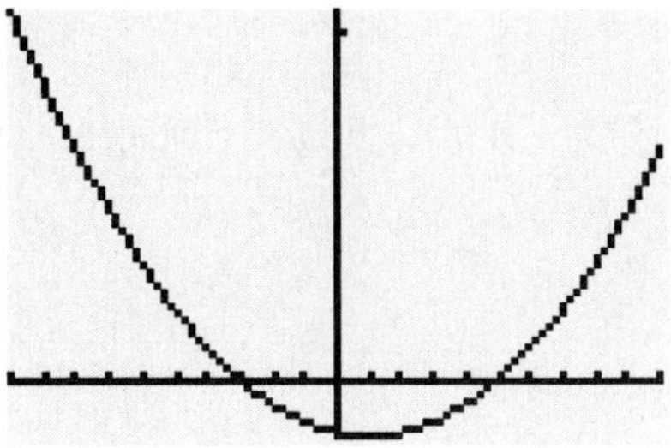

c. graph in part (b).

15. a.

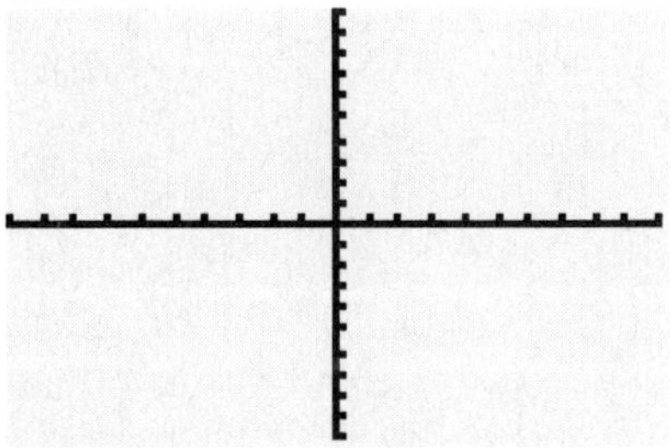

b.

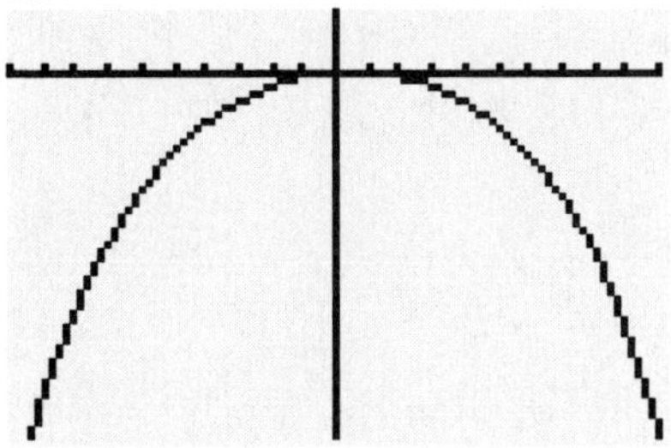

c. graph in part (b).

16. a.

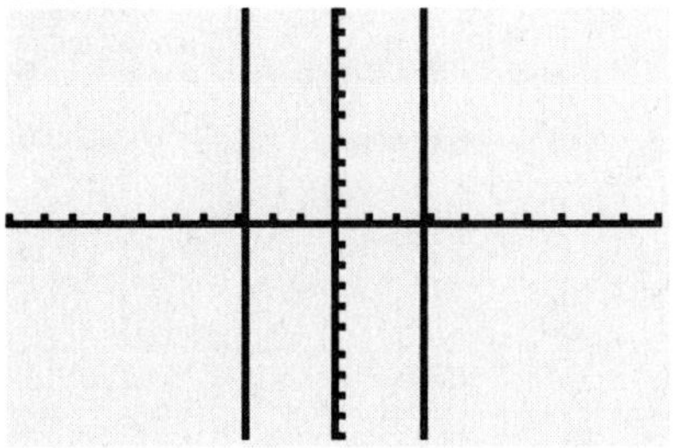

b.

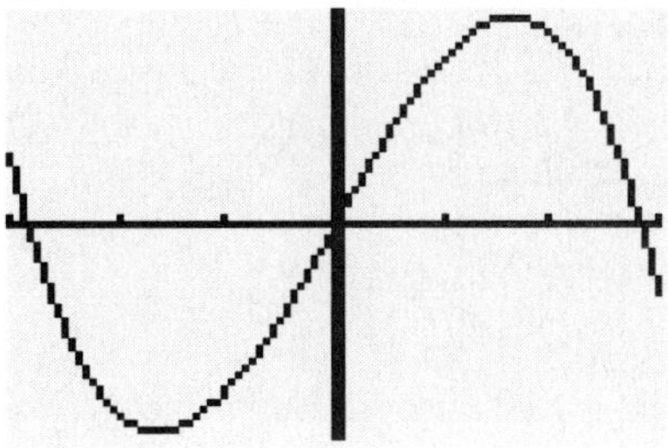

c. graph in part (b).

17.

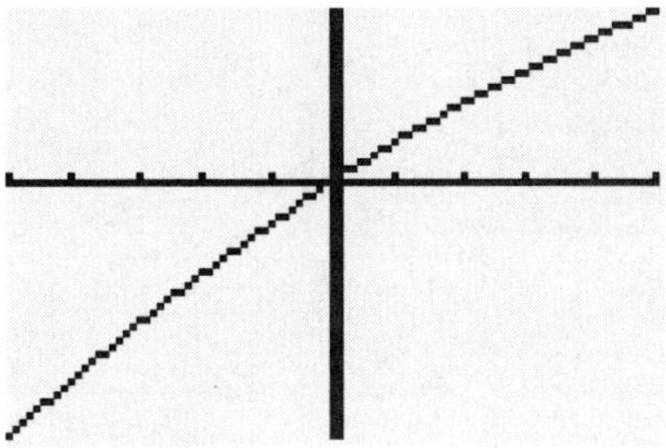

[-150,100]

18.

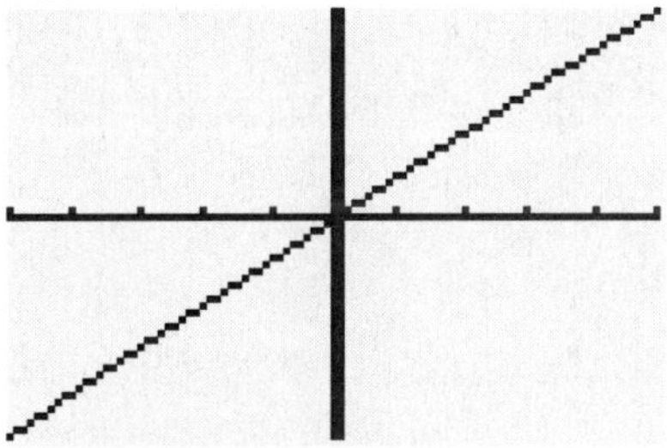

[-102500, 97500]

19.

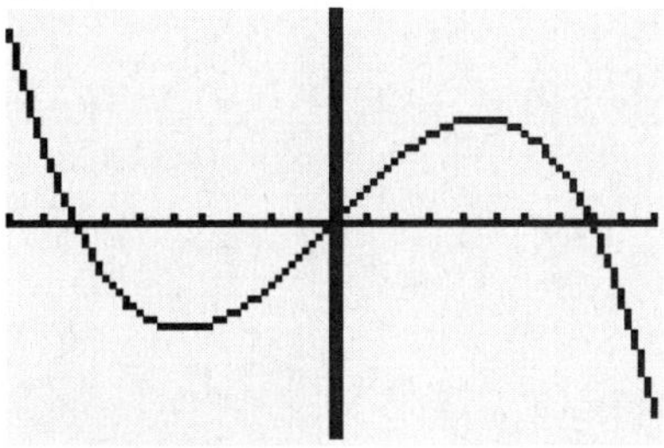

[-4000, 4000]

20.

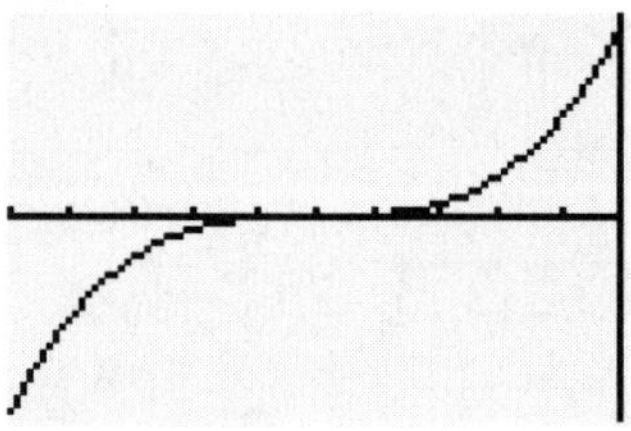

[-130, 130]

21.

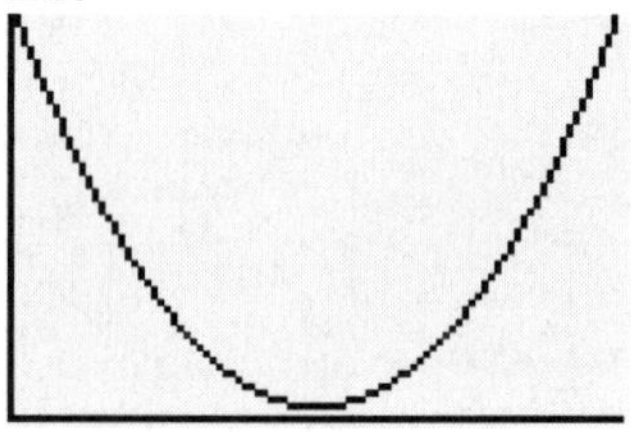

[1200, 11200]

22.

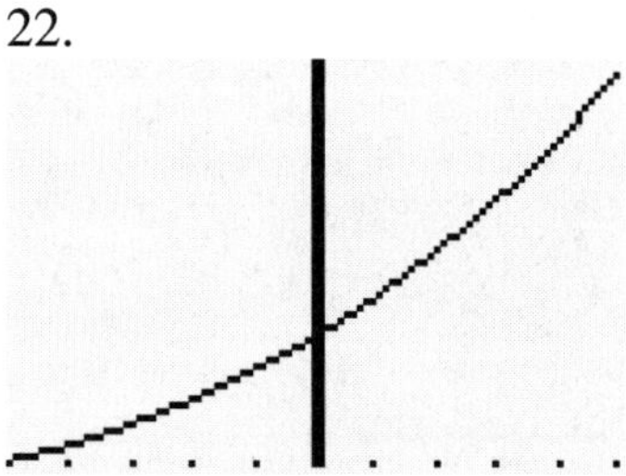

[49,000, 400000]

23.

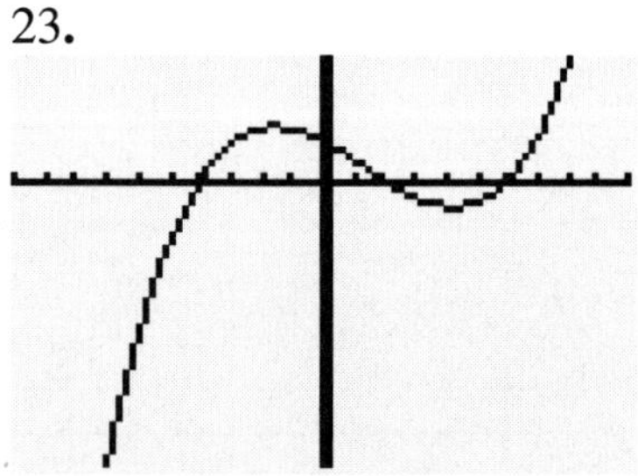

[-10, 10] x [-350, 150]

24.

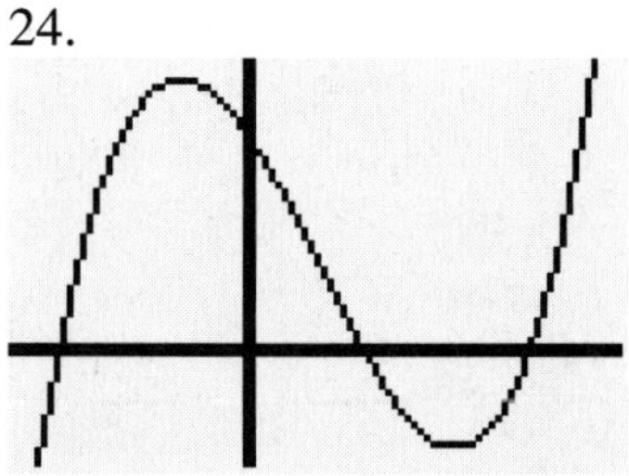

[-50, 80] x [-30000, 80000]

25.

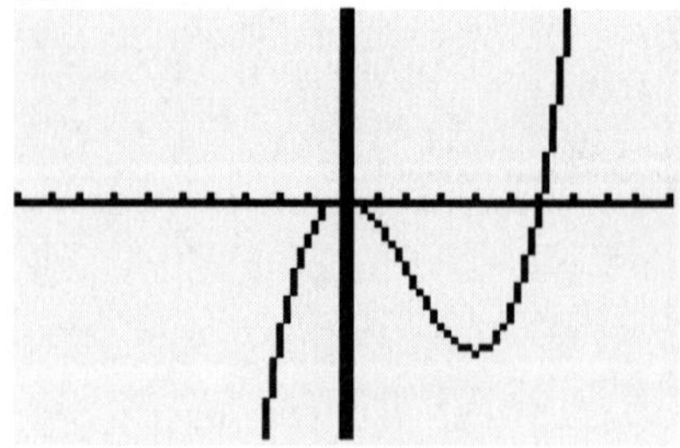

[-10, 10] x [-100, 80]

26.

t	$S(t) = 4.2t + 5.3$
15	68.3
17	76.7
19	85.1
21	93.5

27.

t	$S(t) = 14t^2 - 6t + 2$
20	5,482
30	12,422
40	22,162
50	34,702

28.

29.

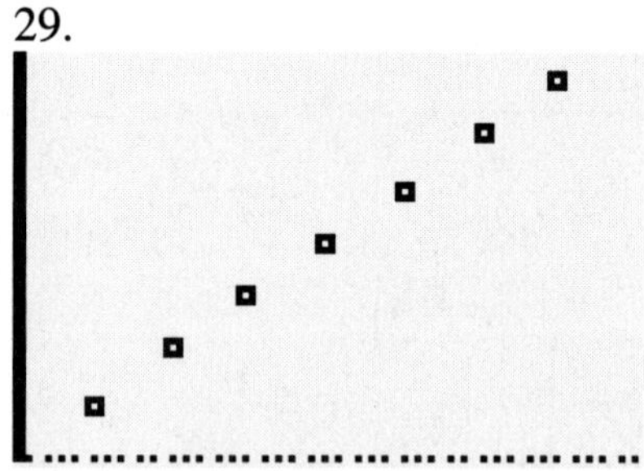

30.

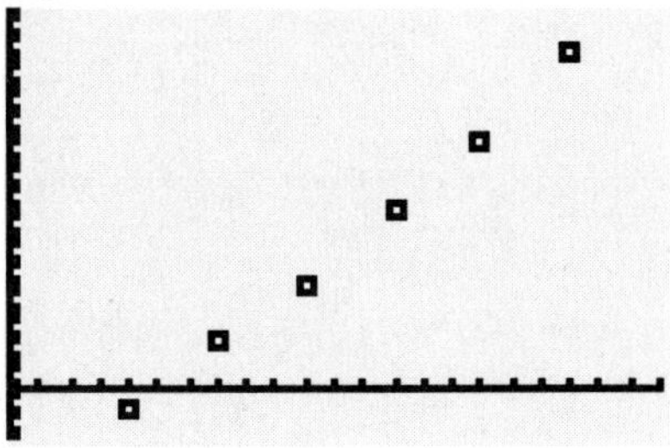

31.

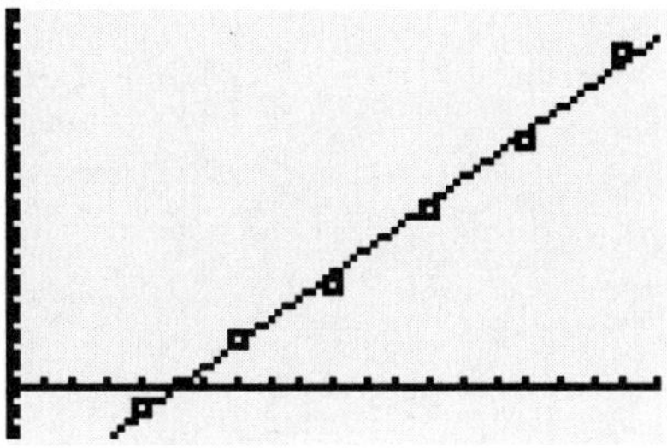

32. yes

33. The x and y values correspond approximately to the relationship $y = 2.49x - 12.75$.

34.

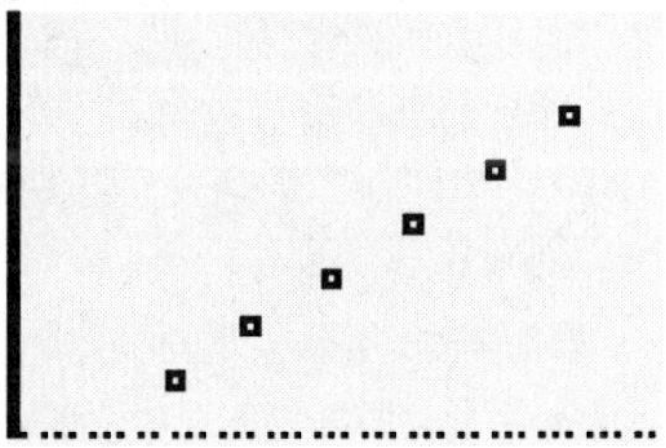

35.

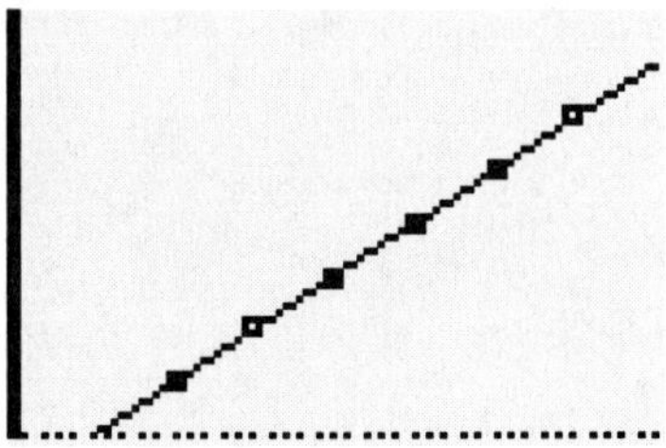

36. yes

37. The x and y values correspond approximately to the relationship $y = 4x + 80$.

38. a. $100 thousand b. 2005

39. a. 125 fish b. 2010

Section 1.3

1. a, b, f	2. -8	3. 5
4. 0	5. 13/5	6. undefined
7. -1	8. 24/19	9. 15

10. -4.5

11. a. positive b. 0 c. negative d. negative

12. $m = 3, b = -7$	13. $m = 2, b = -4.5$
14. $m = -5, b = 11$	15. $m = 1/6, b = -2$
16. $m = -5/3, b = -5$	17. $m = 4, b = 0$
18. $m = 1/3, b = 0$	19. $m = 2, b = -1/2$
20. m = und, b = none	21. $m = -5, b = -3$
22. $m = 0, b = 9$	23. $m = 5, b = -3$

24. a. $m = 2, b = -1$ b. rising graph

25. a. $m = 0.002, b = 0.05$ b. rising

26. a. $m = -100, b = 20{,}000$ b. falling

27. increasing steepness (regardless of rising or falling) 25, 24, 26

28. 7	29. 0.2	30. -5
31. -8	32. 200	33. 75
34. 1.5	35. 1/3	36. -5/4

37. parallel

38. perpendicular

39. perpendicular

40. The rate of change for $f(x) = x$ is 1.

41. The rate of change for $f(x) = 2$ is 0.

Section 1.4

1. $y = 3x - 5$
2. $y = -6x + 4$
3. $y = \frac{2}{3}x$
4. $y = -x + \frac{4}{5}$
5. $y = 6x - 20$
6. $y = -8x - 11$
7. $y = 3x - 1$
8. $y = -2x - 12$
9. $y = 2x - 5$
10. $y = -x - 5$
11. $y = 6$
12. $x = -5$
13. $x = 1$
14. $y = 7$
15. $y = \frac{1}{7}x + \frac{9}{7}$
16. $y = \frac{8}{3}x + 2$
17. $y = \frac{2}{5}x + \frac{7}{5}$
18. $y = 2$
19. $y = -7x - 12$
20. $y = \frac{13}{7}x - \frac{10}{7}$
21. $x = 1$
22. $y = \frac{-2}{3}x - \frac{17}{3}$
23. $y = \frac{1}{3}x + 1$
24. $y = \frac{-7}{4}x + 7$
25. a. $y = 2x + 2$ b. $y = x + 4$ c. $y = 6$
26. $y = -4x + 6$
27. $y = \frac{1}{4}x - 2$
28. slope = 11
29. slope = 5
30. 3
31. -1
32. $2x$
33. $8x$
34. $2x - 4$
35. $6x$
36. $y = 3x + 14$
37. $y = 6x + 70$
38. $y = 0.2x + 40$

CHAPTER 2 ANSWER KEY

Section 2.1

1. x = -4 2. x = 2/5 3. x = 1

4. x = 4 5. x = -8/3 6. x = 5 ½

7. x = -11/12 8. x = 8 9. x = 5

10. x = 4 11. x = -2 12. x = -22

13. a. x = -30 b. (-30, 0) c. x = -30

14. a. x = 4 b. (4, 0) c. x = 4

15. a. x = -6 b. (-6, 0) c. x = -6

16. a. (-3, 0) b. (-3, 0)

17. a. (8, 0) b. (8, 0)

18. a. (2, 0) b. (0, 4) c. x = 2

19. a. (3, 0) b. (0, -15) c. x = 3

20. a. (-2, 0) b. (0, -3) c. x = -2

21. a. (-20, 0) b. (0, 2) c. x = -20

22. $y = \dfrac{2t - ax}{t}$ 23. $h = \dfrac{2A}{b_1 + b_2}$

24. $x = \dfrac{d - by}{a - c}$ 25. $x = \dfrac{2a - 2y}{3 - a}$

26.

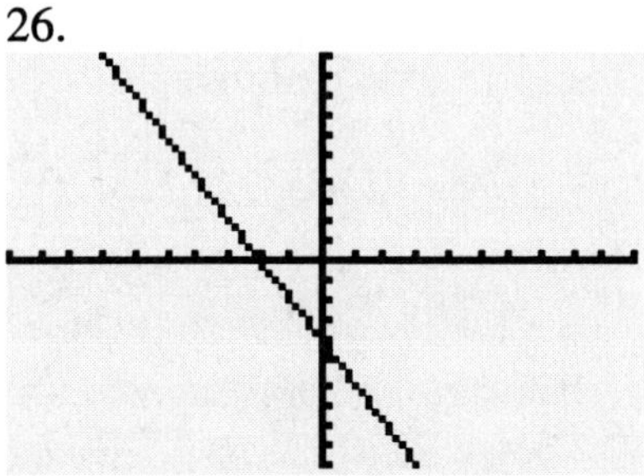

27.

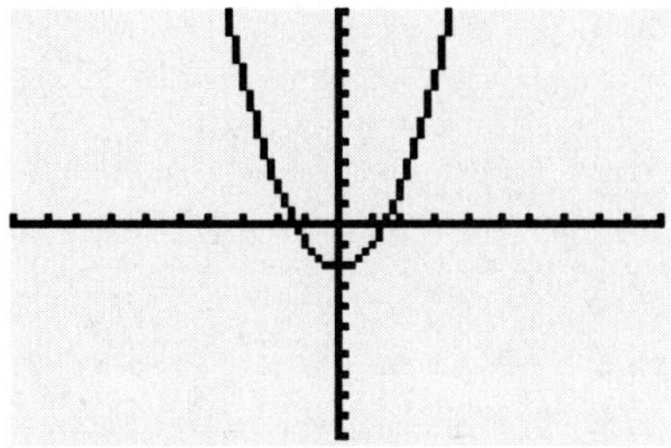

28.

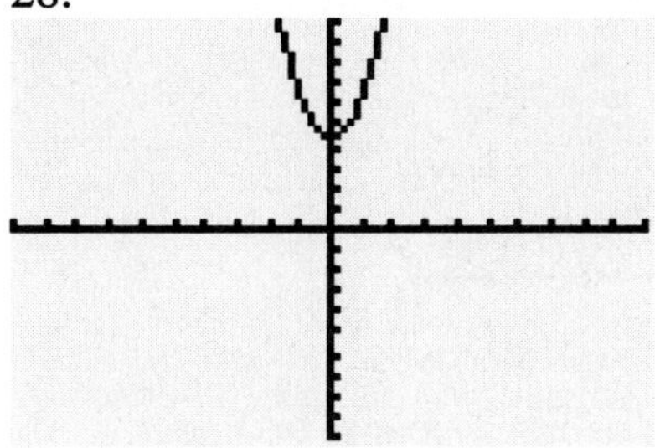

29.

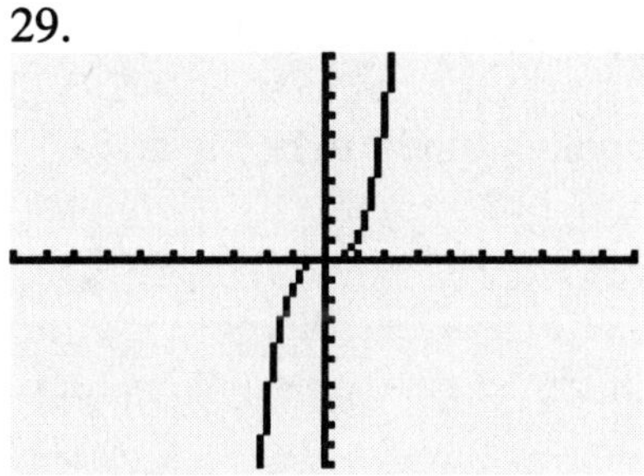

Section 2.2

1. yes, linear

2. no, nonlinear

3.

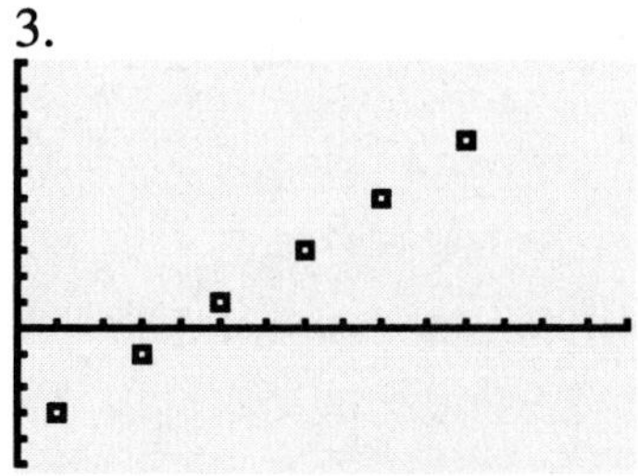

4.

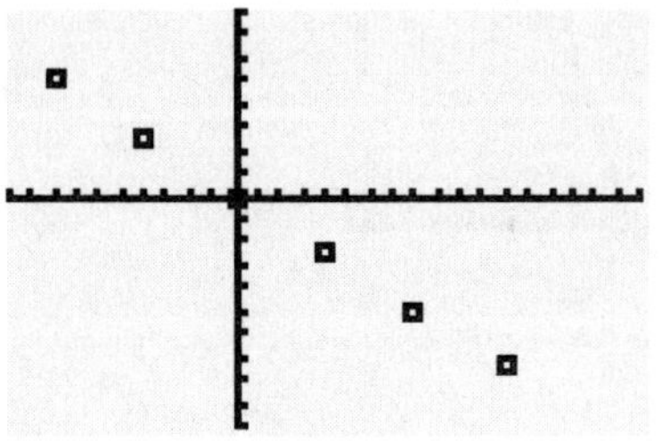

5.

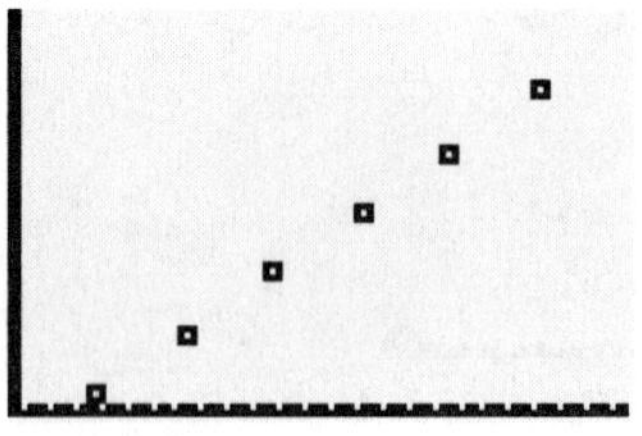

6. Yes, the data points appear to be linear.

7. Yes, the data points appear to be linear.

8. Yes, the data points appear to be linear.

9. $f(x) = x - 4$

10. $g(x) = -0.625x$

11. $h(x) = 1.5x - 10$

12. $f(10) = 6, f(-2.5) = -6.5$

13. $g(15) = -9.375, g(-10) = -6.25$

14. $h(1.5) = -7.75, h(-7) = -20.5$

15.

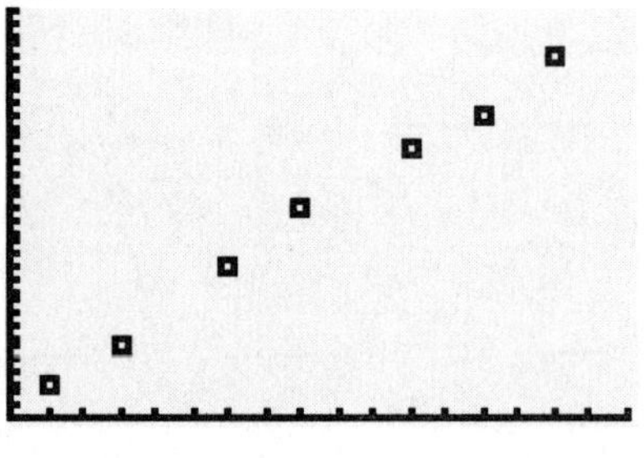

16. Yes, the data points appear to be linear.

17. $f(x) = 2.001x + 0.850$

18. $f(7) = 14.857$

19.

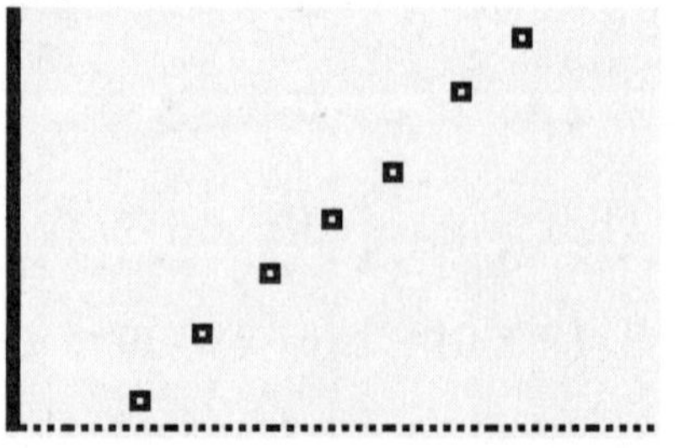

20. Yes, the data points appear to be linear.

21. $f(x) = 4.029x - 0.571$

22. $f(7) = 27.632$

23. $y = 3x + 1$

24. $y = 2.5x - 2$

25. a. approximately linear
b. nonlinear c. exactly linear

26. b

27. d

28. c

29. a

30. e

Section 2.3

1. (1, 5)

2. (-2, 1)

3. (0, 4)

4. no solution

5. (3, 4)

6. (0.5, -2.5)

7. (-1, 4)

8. (0, -3)

9. (6, 1)

10. (3, 2)

11. (-4, 2)

12. (-2, 1)

13. No solution, the lines are parallel.

14. Many solutions, they are the same line.

15. (5, 2)

16. (3, -4)

17. (3, 3)
18. (6, -2)
19. (5, -7)
20. (0.5, 5)
21. (4, 0)
22. (-2, -2)
23. (1, -1)
24. (1.5, 4)
25. (-5, 2)
26. (4, -1/3)
27. (4, 2)
28. no solution
29. (2/3, 5)
30. $\{(x, y): y = -\frac{3}{2}x + 7\}$

Section 2.4

1. $x \leq 4$
2. $x < -4.5$
3. $x > -2$
4. $x \leq 9$
5. $x < 7/3$
6. $x > 11/4$
7. $x \geq 3$
8. $x < 29/16$
9. $x > 1$
10. $x \geq -\frac{7}{8}$
11. all real numbers
12. $x < -5/3$
13. $x > 5$
14. $x > -23/11$
15. no solution
16. $x \leq -29$
17. $x > -11$
18. $x > 38/13$
19. $x \geq 7$
20. $x < 6$
21. $x \leq 2/3$
22. $x < 1$
23. $-3 < x < 4$
24. $-3 \leq x \leq 3$
25. $5 > x > 3$
26. $1 < x \leq 4$
27. $x > 4$
28. no solution
29. $x < -1$
30. $x \geq 2$
31. $x > 3$
32. $x < 12$

Section 3.1

1. a. quadratic b. concave up
c. minimum point

2. a. quadratic b. concave down
c. maximum point

3. a. not quadratic

4. a. quadratic b. concave up
c. minimum point

5. a. not quadratic

6. a. quadratic b. concave up
c. minimum point

7. a.

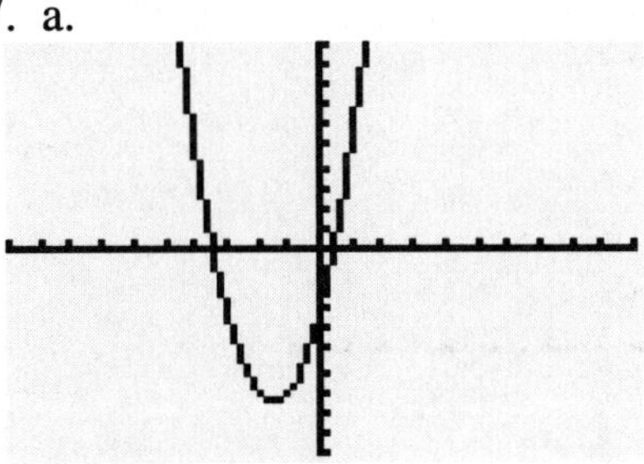

b. complete graph

8. a.

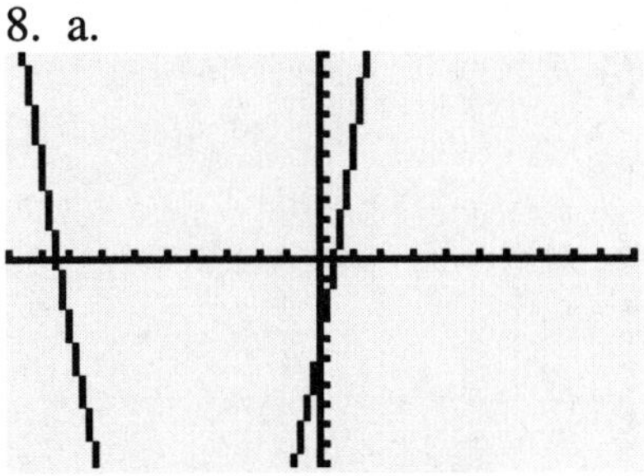

b. not a complete graph

9. a.

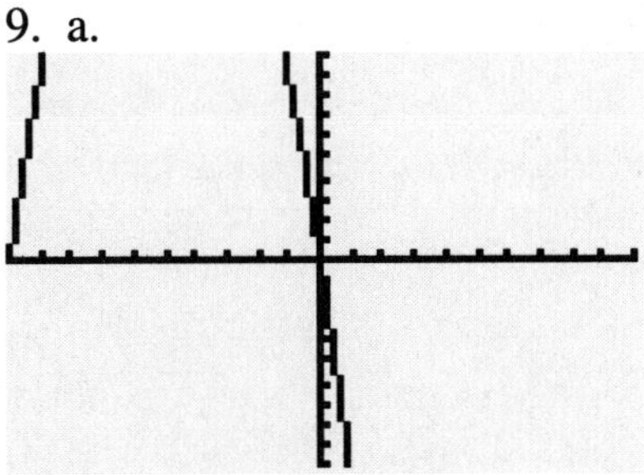

b. not a complete graph

CHAPTER 3 ANSWER KEY

10. a.

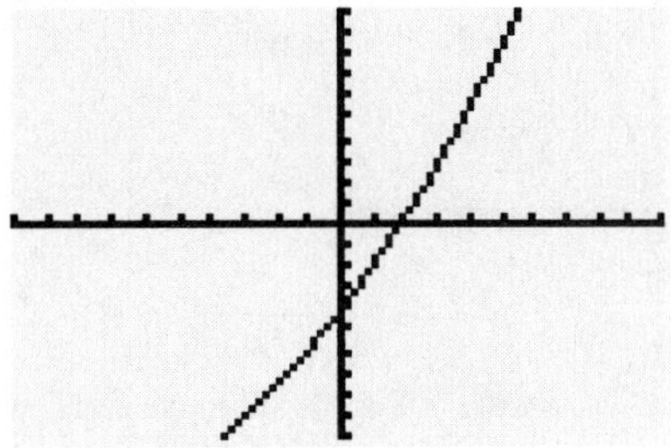

b. not a complete graph

11. a.

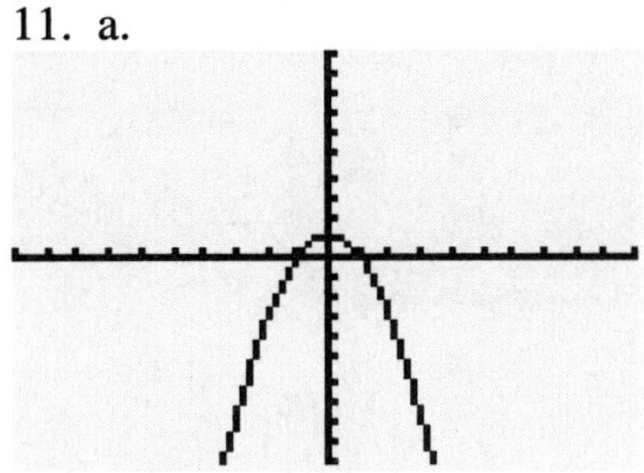

b. complete graph

12. a.

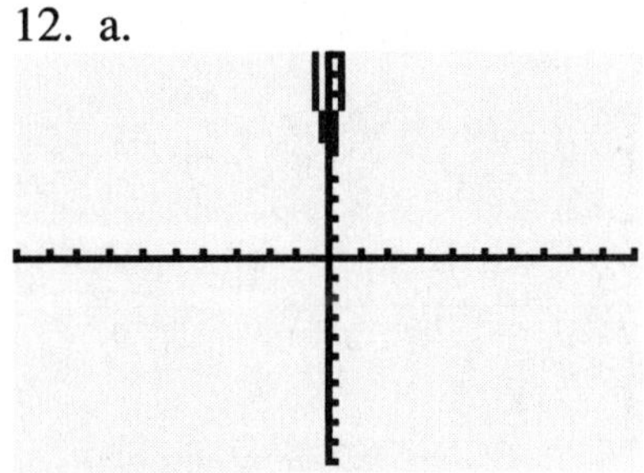

b. complete graph

13. a.

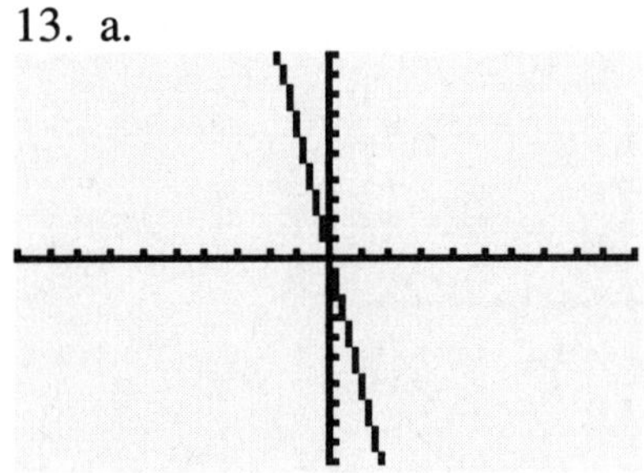

b. not a complete graph

14. a.

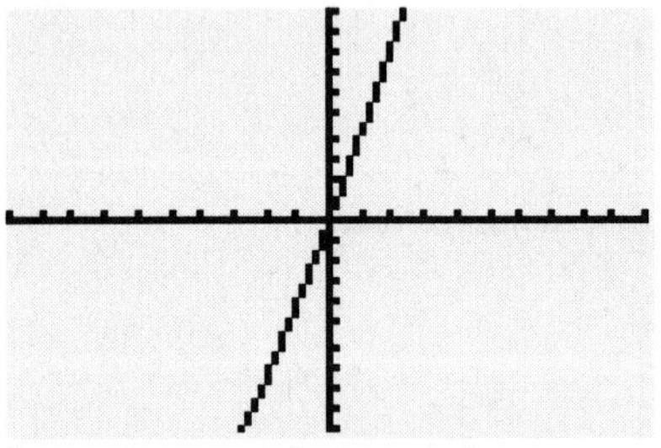

b. not a complete graph

15. a. (2, 3)
b. [-10, 10] x [-10, 10]

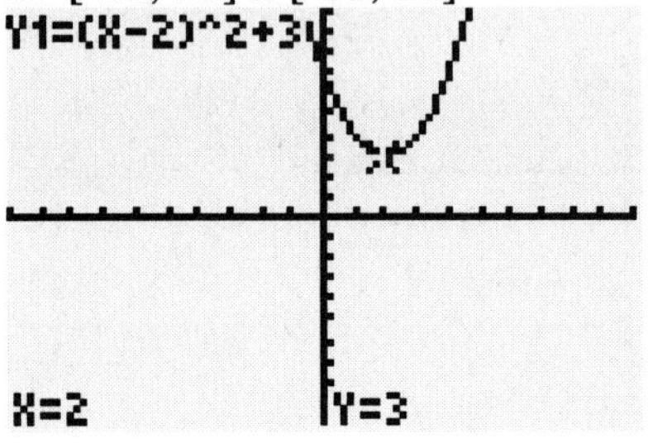

16. a. (-1, -3)
b. [-10, 10] x [-10, 10]

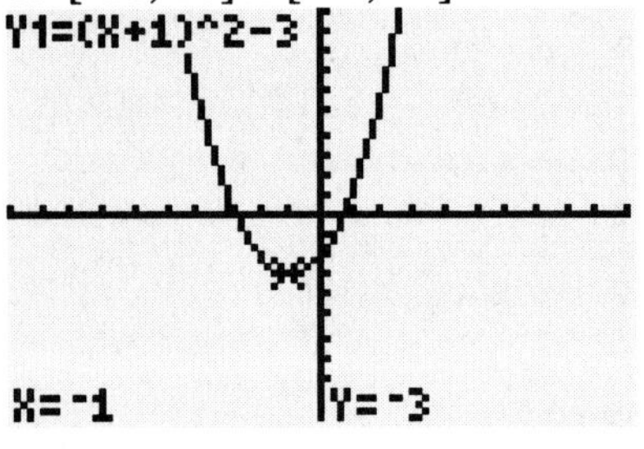

17. a. (1, 3)
b. [-10, 10] x [-10, 10]

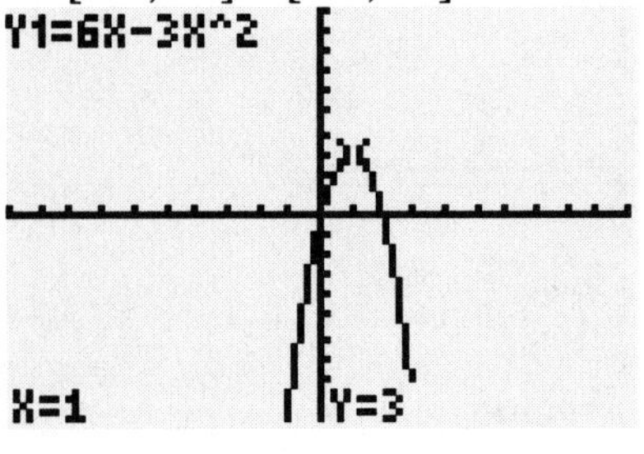

18. a. (-2, 2)
b. [-10, 10] x [-10, 10]

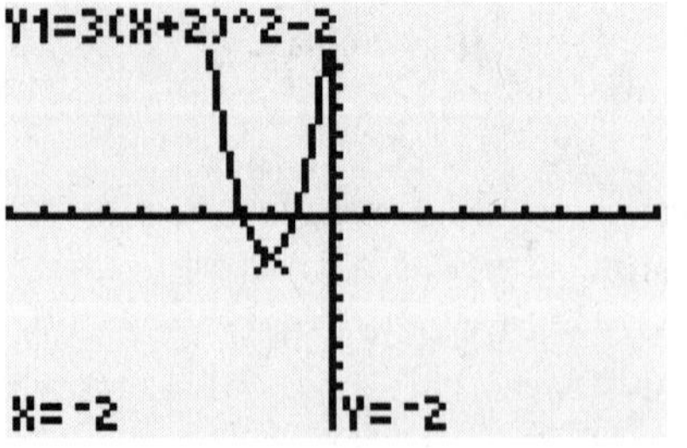

19. a. (-4, 2)
b. [-10, 10] x [-10, 10]

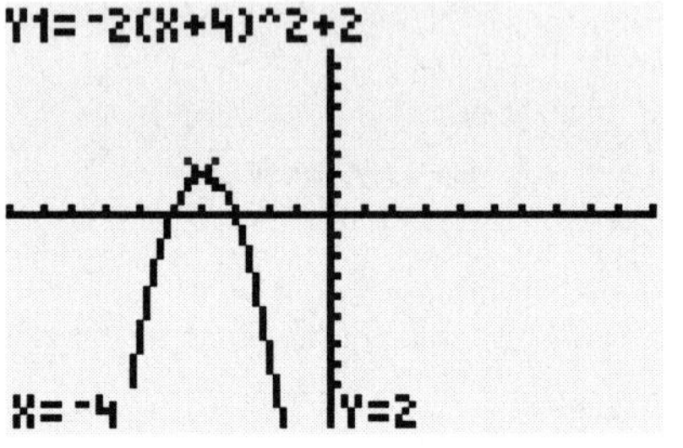

20. a. (-3, 4)
b. [-10, 10] x [-20, 5]

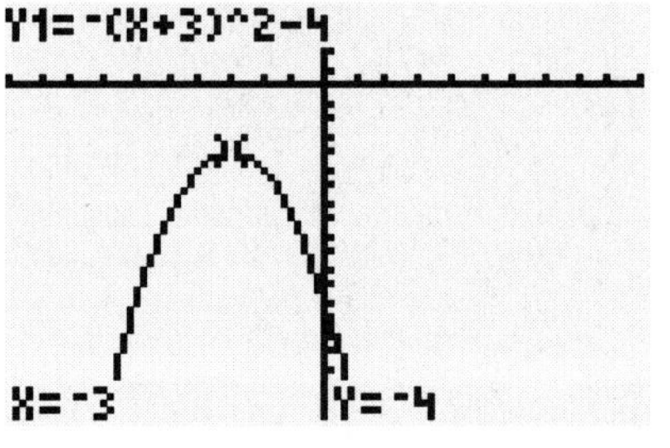

21. a. (0, 7)
b. [-10, 10] x [-5, 2]

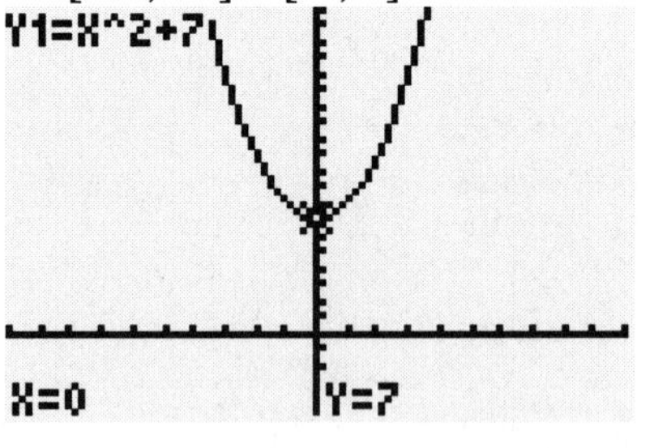

22. a. (-1, -2)
b. [-10, 10] x [-5, 20]

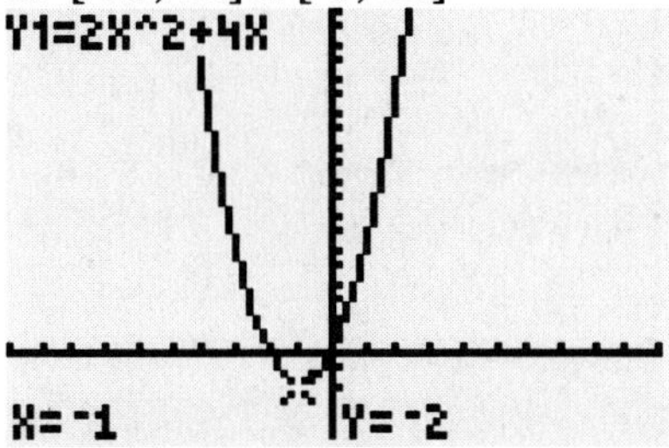

23. a. (1, -1)
b. [-10, 10] x [-5, 20]

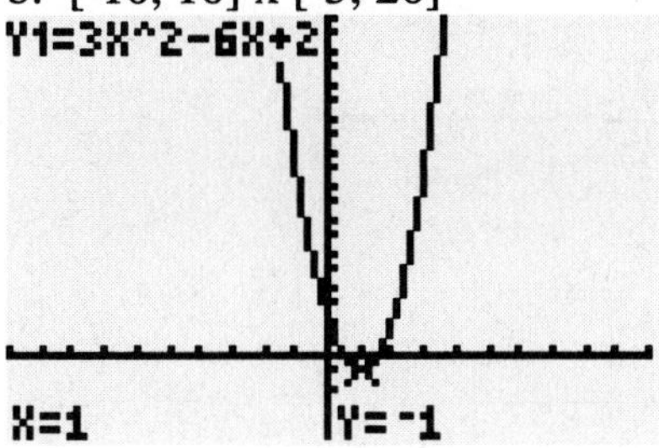

24.

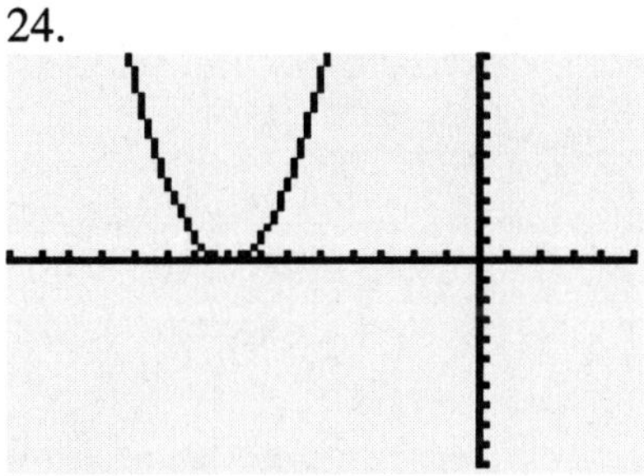

[-15. 5] x [-10, 10]

25.

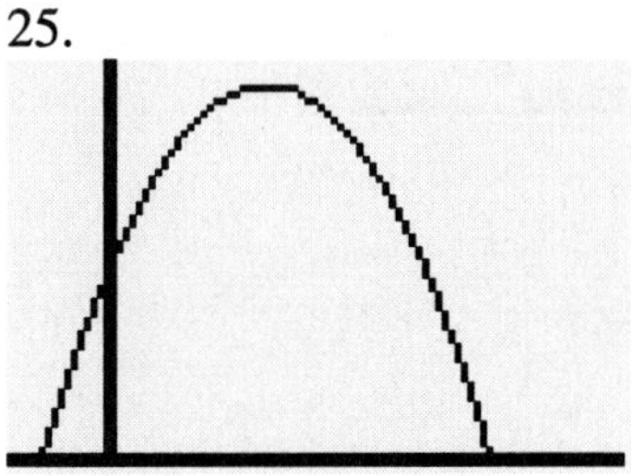

[-15, 80] x [-10, 1300]

26.

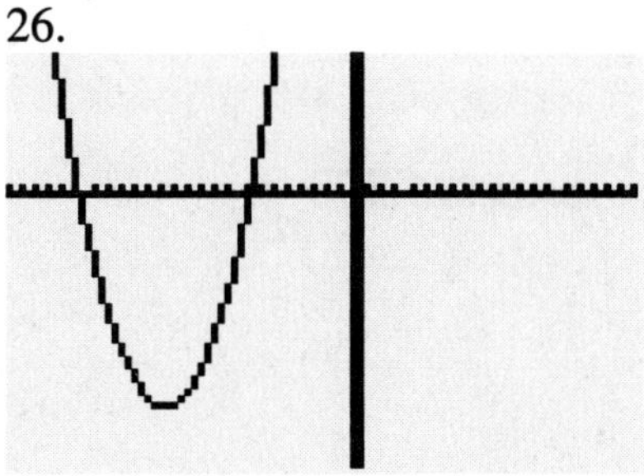

[-25, 20] x [-100, 50]

27. (6, 0) and (-15, 0)

28. (-3, 0) and (2.5, 0)

29. (-0.8, 0) and (8, 0)

Section 3.2

1. x = -3, 1
2. x = 5, -4
3. x = 2, -6
4. x = -3/2, 5
5. x = -1/3, -4.5
6. x = -4/3, ½
7. x = -2/5, -8
8. x = -3/2
9. x = -1/6, 4
10. (-4, 0) and (2, 0)
11. (-10, 0) and (-3, 0)
12. (1/2, 0) and (6, 0)
13. (1/2, 0) and (5/3, 0)
14. (-3/4, 0) and (3/2, 0)
15. (1/3, 0) and (4, 0)
16. x = 40, -30
17. x = -15, 40
18. x = 15/2, -50
19. x = -5/2, 40
20. x = -3/10, ¾
21. x = -12/15, -10
22. x = 3, -3
23. $x = 2\sqrt{7}, -2\sqrt{7}$
24. $x = 2\sqrt{3}, -2\sqrt{3}$
25. x = 6, 2
26. x = -3/2, -4
27. x = 10, -7
28. x = -1/2, 4/3

29. $x=\dfrac{2+\sqrt{10}}{6},\dfrac{2-\sqrt{10}}{6}$

30. $x=\dfrac{7+\sqrt{11}}{2},\dfrac{7-\sqrt{11}}{2}$

31. $x=5+\sqrt{7},5-\sqrt{7}$

32. $x=1+\sqrt{3},1-\sqrt{3}$

33. $x=\dfrac{-3+\sqrt{41}}{4},\dfrac{-3-\sqrt{41}}{4}$

34. $x=2+i,2-i$

35. $x=\sqrt{35}\,i,-\sqrt{35}\,i$

36. $x=3+\sqrt{5}\,i,3-\sqrt{5}\,i$

37. $x=\dfrac{-4+\sqrt{38}\,i}{2},\dfrac{-4-\sqrt{38}\,i}{2}$

38. $x=\dfrac{1+\sqrt{6}\,i}{2},\dfrac{1-\sqrt{6}\,i}{2}$

39. $x=\dfrac{1+2i}{3},\dfrac{1-2i}{3}$

Section 3.3

1.

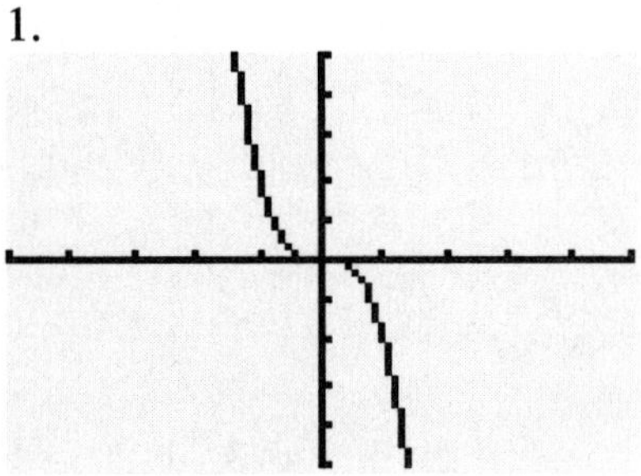

2.

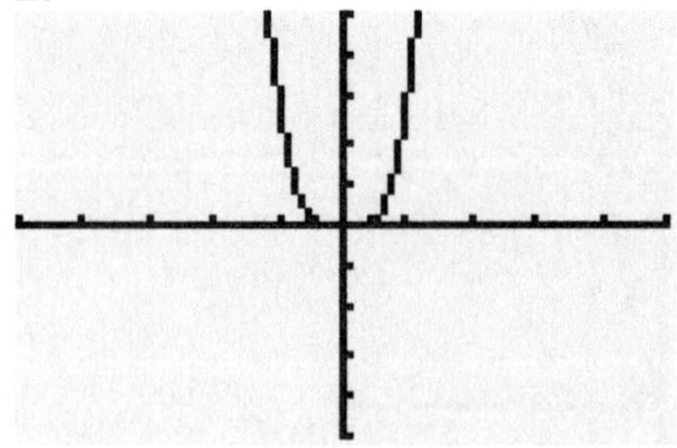

3.

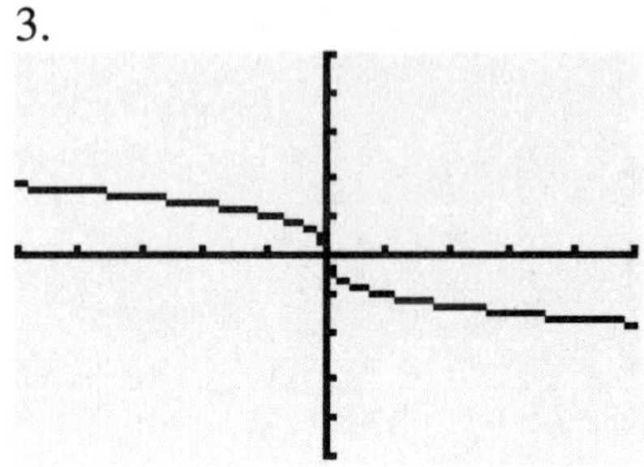

4.

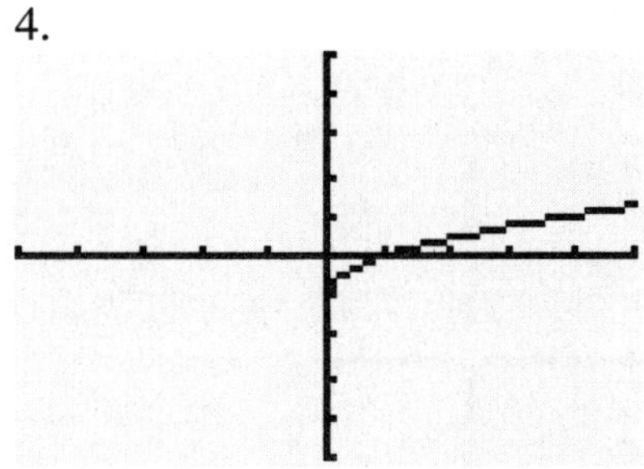

5.

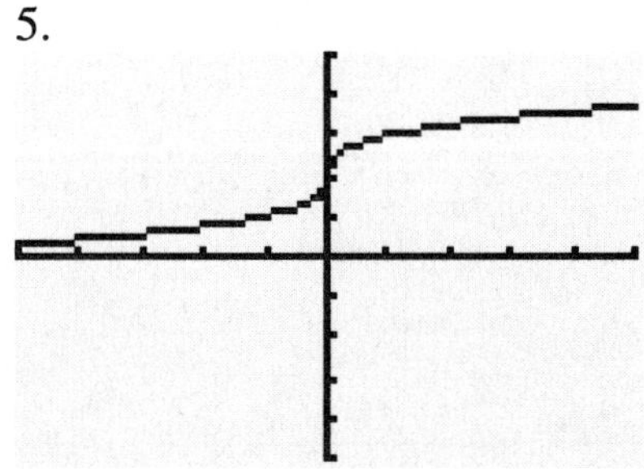

6.

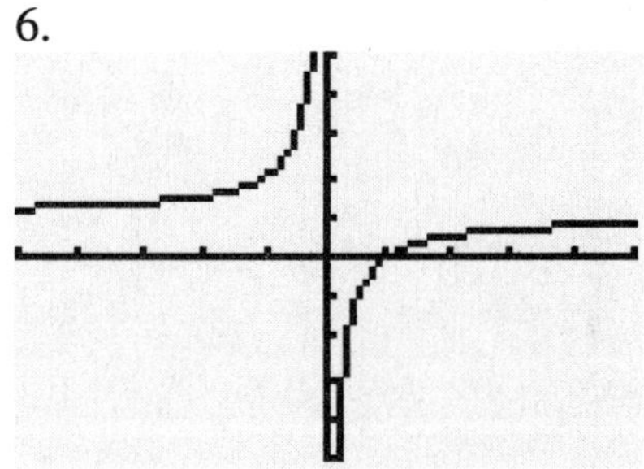

7.

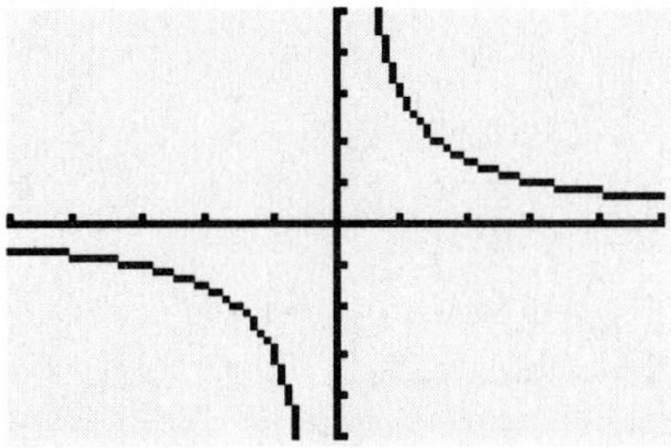

8.

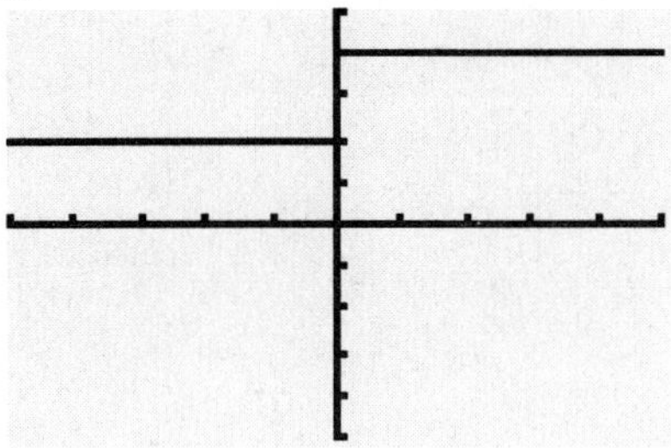

9.

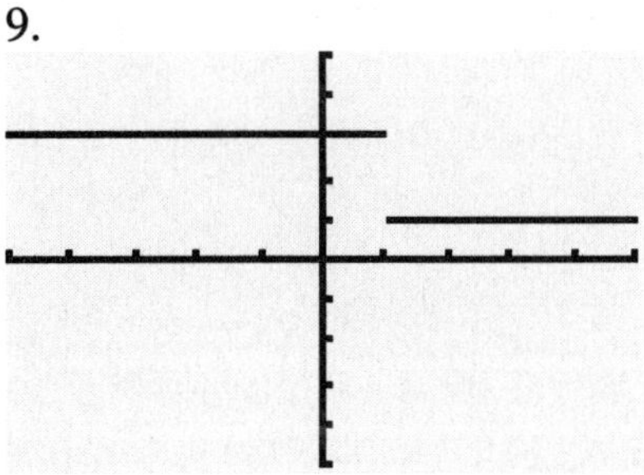

10.

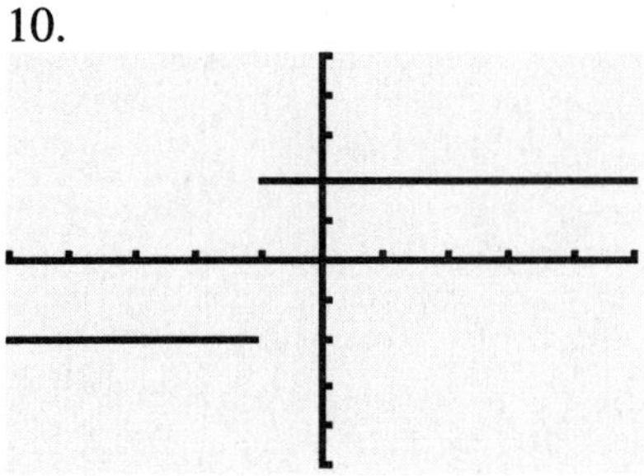

11. a.

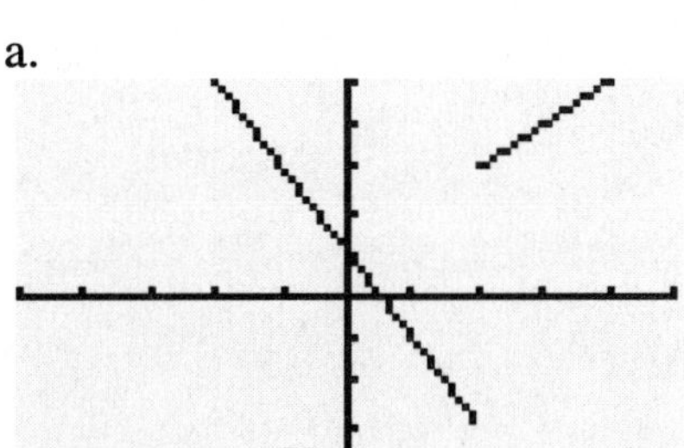

b. $f(1) = -1,\ f(5) = 6$

12. a.

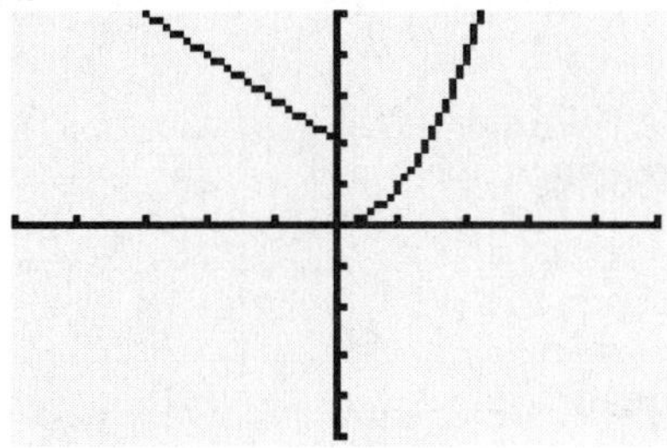

b. $f(-2) = 4,\ f(3) = 9$

13. a.

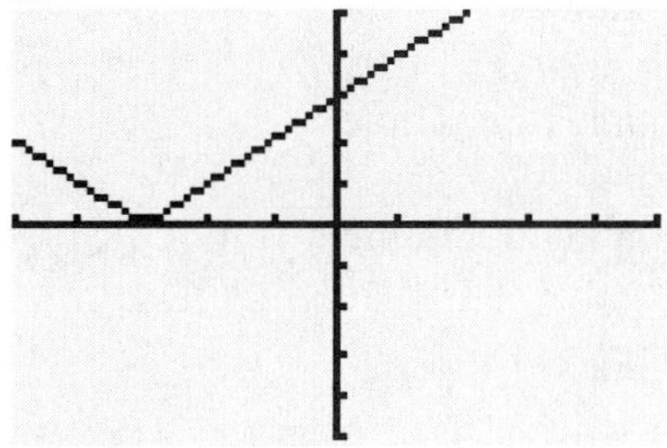

b. $f(-4) = 1,\ f(6) = 9$

14. a.

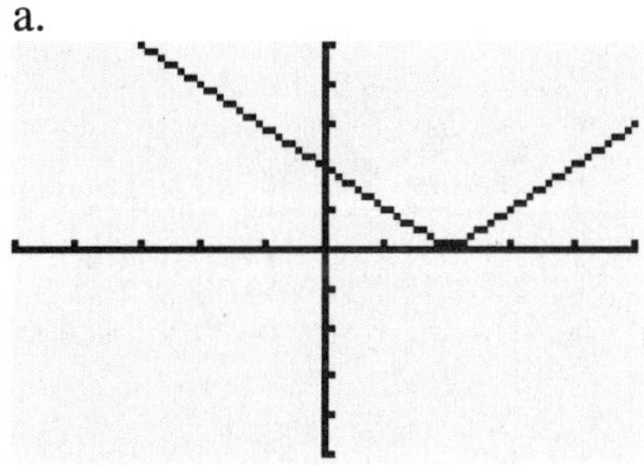

b. $f(4) = 2,\ f(8) = 6$

15. $f(-1) = -16,\ f(2) = 2$

16. $f(-1) = 7,\ f(2) = 13$

17. $f(-1) =$ Does Not Exist, $f(2) = 0$

18. $f(-1) = -1,\ f(2) = \sqrt[3]{2}$

19. $f(-1) =$ Does Not Exist, $f(2) = 1/3$

20. $f(-1) = 3/2,\ f(2) = 2.25$

21. $f(-1) = 4,\ f(2) = 8$

22. $f(-1) = 3,\ f(2) = 4$

23. $f(-1)=3,\ f(2)=4$

24. a. decreasing
    b. decreasing

Section 3.4

1. exactly quadratic

2. not quadratic

3. exactly quadratic

4. approximately quadratic

5. approximately quadratic

6. $y=2x^2+3x-4$

7. $y=-x^2+4x+1$

8. a.

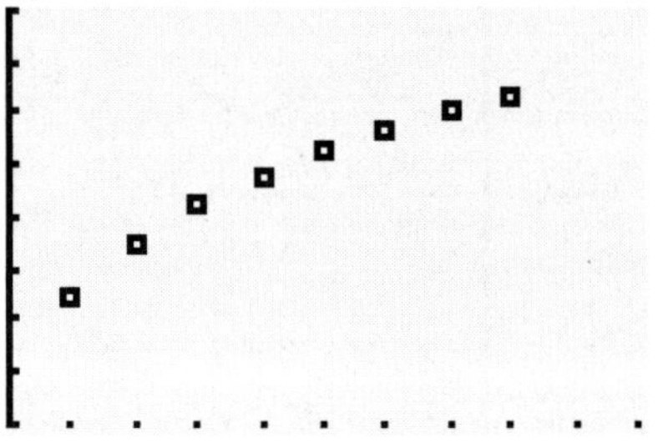

b. $y=6.4x^{0.23}$
c. $y=0.533x+6.378$
d. power function

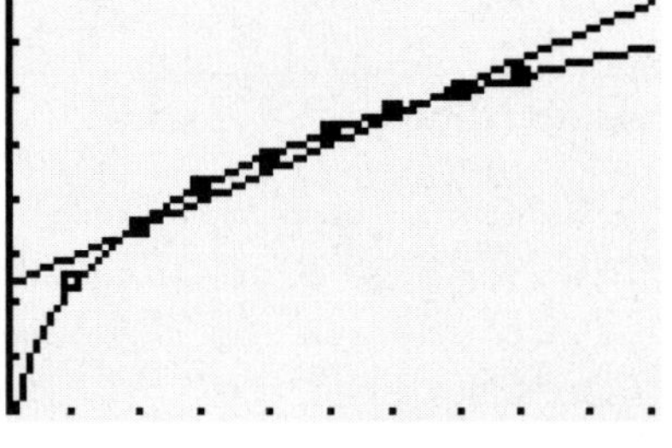

9. $y=x^{1.2}$

10. $y=0.1x^{2.5}$

11.

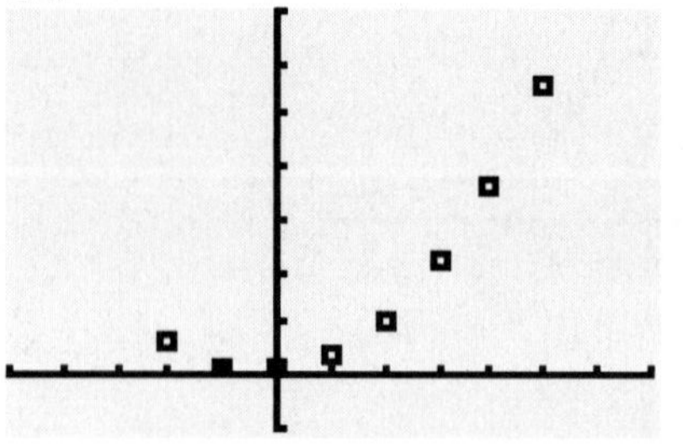

12. $y=0.7x+0.7$
$y=0.2x^2+0.1x+0.1$

13. quadratic is a better fit

14.

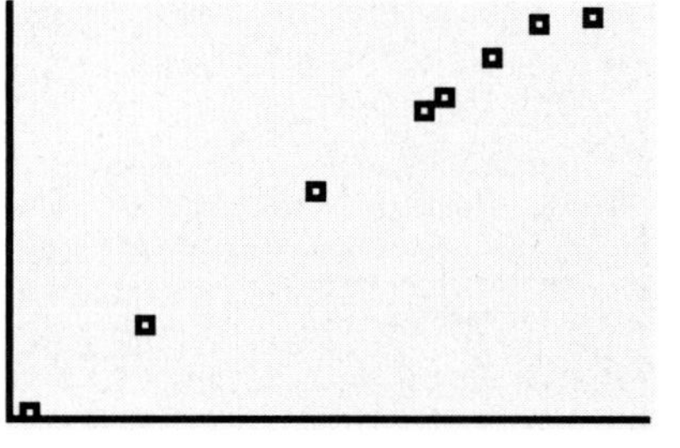

15. $y=491.3x+3275.3$
$y=-0.7x^2+730.2x-15176.0$

16. linear is a better fit

CHAPTER 4 ANSWER KEY

Section 4.1

1.

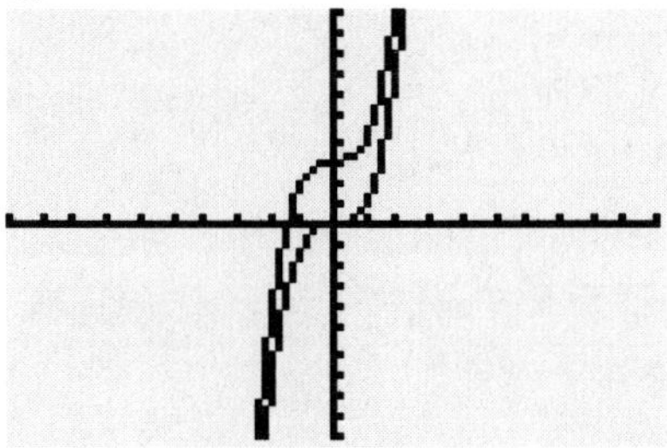

2.

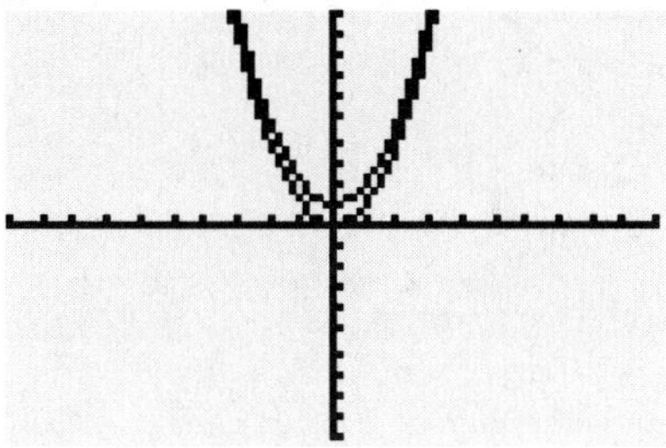

3.

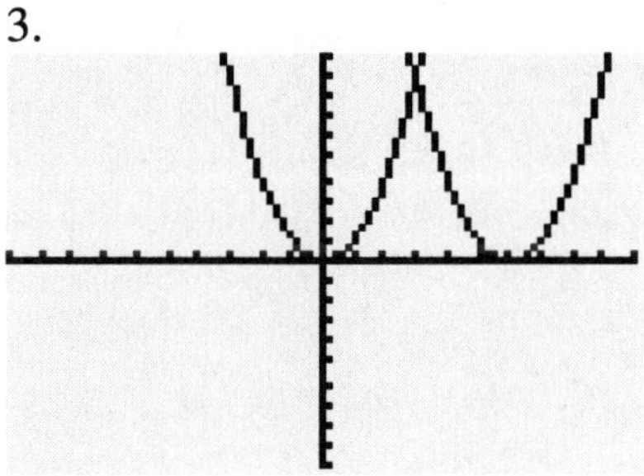

4.

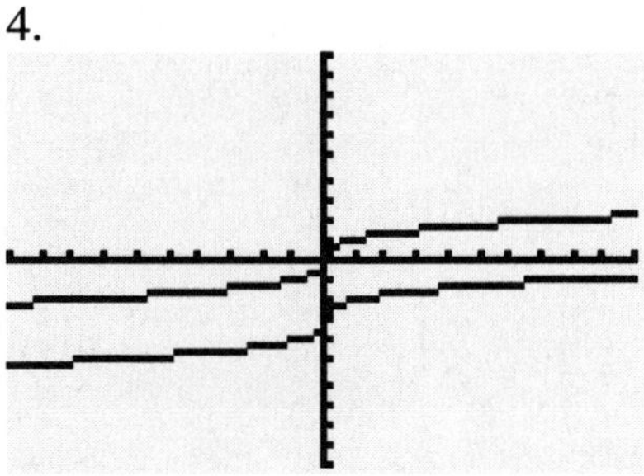

5.

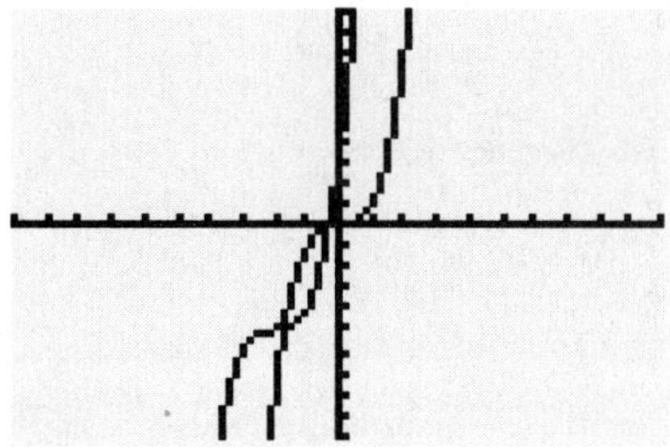

6.

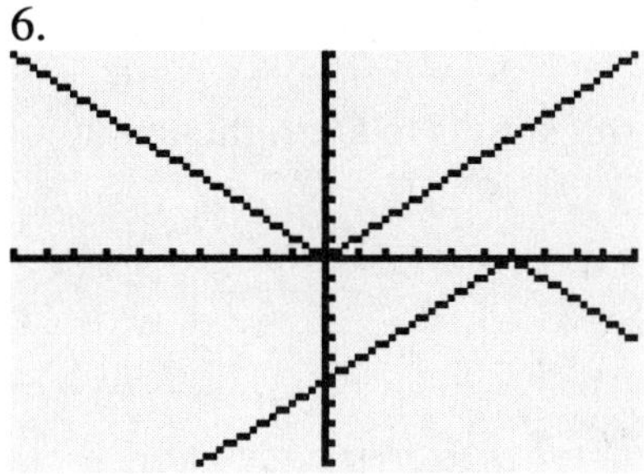

7.

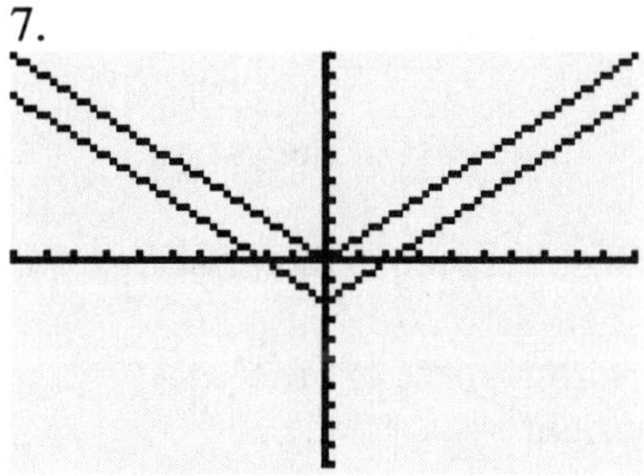

8.

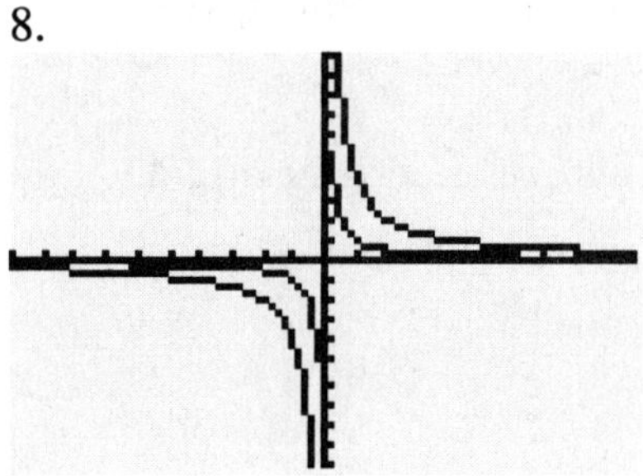

9.

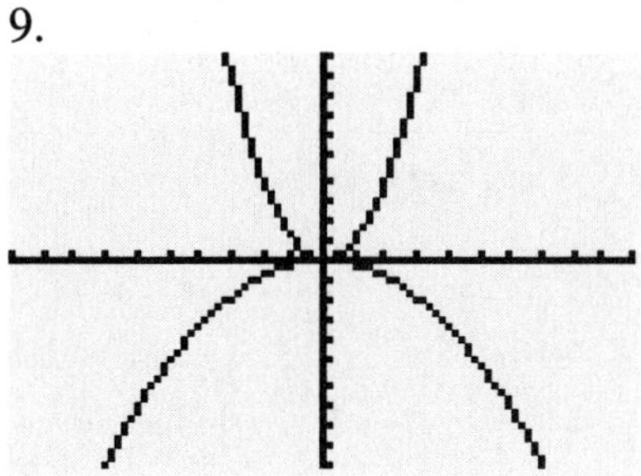

10.

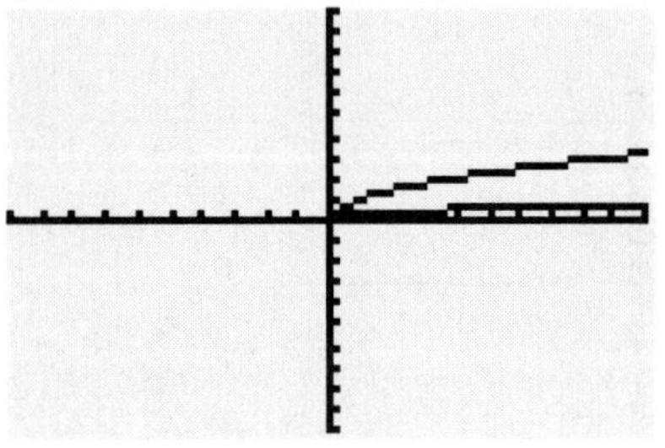

11. Shift the graph 1 unit to the left and down 3 units.

12. Shift the graph 4 units to the right and up 6 units.

13. $y = \sqrt{x+4} + 2$

14. $y = (x-1)^{1/3} - 3$

15. Symmetric with respect to the *y*-axis.

16. Symmetric with respect to the *y*-axis.

17. Symmetric with respect to the origin.

18. Symmetric with respect to the *x*-axis, *y*-axis, and the origin.

19. Symmetric with respect to the *x*-axis, and the *y*-axis.

20. Symmetric with respect to the origin.

21. even

22. odd

23. neither

Section 4.2

1. a. $(f+g)(x) = 3x+3$
b. $(f-g)(x) = x+11$
c. $(f \cdot g)(x) = 2x^2 - x - 28$
d. $\left(\frac{f}{g}\right)(x) = \frac{2x+7}{x-4}$
e. Domain = $\{x : x \neq 4\}$

2. a. $(f+g)(x) = 2x^2 + 2$
b. $(f-g)(x) = 10$
c. $(f \cdot g)(x) = x^4 + 2x^2 - 24$
d. $\left(\frac{f}{g}\right)(x) = \frac{x^2+6}{x^2-4}$
e. Domain = $\{x : x \neq 2, -2\}$

3. a. $(f+g)(x) = x^2 + 5x - 5$
b. $(f-g)(x) = x^2 - x + 5$
c. $(f \cdot g)(x) = 3x^3 + x^2 - 10x$
d. $\left(\frac{f}{g}\right)(x) = \frac{x^2+2x}{3x-5}$
e. Domain = $\{x : x \neq \frac{5}{3}\}$

4. a. $(f+g)(x) = \frac{5x+2}{x(x+2)}$
b. $(f-g)(x) = \frac{-3x+2}{x(x-2)}$
c. $(f \cdot g)(x) = \frac{4}{x(x+2)}$
d. $\left(\frac{f}{g}\right)(x) = \frac{x+2}{4x}$
e. Domain = $\{x : x \neq 0\}$

5. a. $(f+g)(x) = \sqrt{x} + x^2 - 6x$
b. $(f-g)(x) = \sqrt{x} - x^2 + 6x$
c. $(f \cdot g)(x) = \sqrt{x}(x^2 - 6x)$
d. $\left(\frac{f}{g}\right)(x) = \frac{\sqrt{x}}{x^2-6x}$
e. Domain = $\{x : x \neq 0, 6\}$

6. a. $(f+g)(x) = \frac{4x+3}{6}$
b. $(f-g)(x) = \frac{5}{6}$
c. $(f \cdot g)(x) = \frac{2x^2+3x-2}{18}$

d. $\left(\frac{f}{g}\right)(x)=\frac{2(x+2)}{2x-1}$

e. Domain = $\{x : x \neq \frac{1}{2}\}$

7. a. $(f+g)(2)=14$
b. $(f-g)(0)=-7$
c. $(f\cdot g)(3)=-180$
d. $\left(\frac{f}{g}\right)(-4)=\frac{-13}{4}$

8. a. $(f+g)(3)=2.5$
b. $(f-g)(0)=-1$
c. $(f\cdot g)(8)=\frac{2}{3}$
d. $\left(\frac{f}{g}\right)(15)=32$

9. $(f\circ g)(x)=3x+9$
$(g\circ f)(x)=3x+11$

10. $(f\circ g)(x)=2x-1$
$(g\circ f)(x)=2x-5$

11. $(f\circ g)(x)=4(x+9)^2$
$(g\circ f)(x)=4x^2+9$

12. $(f\circ g)(x)=\frac{1}{6x^2}$
$(g\circ f)(x)=\frac{6}{x^2}$

13. $(f\circ g)(x)=|4x+5|$
$(g\circ f)(x)=4|x+3|+2$

14. $(f\circ g)(x)=\sqrt{3x^2-5}$
$(g\circ f)(x)=3x-15$

15. a. 7 b. 55
c. 49 d. 127

16. a. 3 b. -4
c. 2 d. -1
e. 4 f. 2

Section 4.3

1. {(4, 1), (8, 2), (6, 3), (10, 4)}

2. No inverse

3. {(4, 4), (5, 5), (6, 6)}

4. {(0, -1), (2, 1), 3, 2)}

5. a. $f(g(x))=\frac{1}{x},\ g(f(x)=\frac{36}{x}$
b. No, not inverses

6. a. $f(g(x))=x,\ g(f(x)=x$
b. Yes, inverses

7. a. $f(g(x))=\frac{3x-16}{3},\ g(f(x)=x-16$
b. No, not inverses

8. a.
$f(g(x))=\sqrt{(x-5)^2+5},\ g(f(x)=(\sqrt{x+5}-5)^2$
b. No, not inverses

9. a. $f(g(x))=x,\ g(f(x)=x$
b. Yes, inverses

10.

x	$f(x)$
-2	9
-1	7
0	5
1	3

x	$f^{-1}(x)$
9	-2
7	-1
5	0
3	1

11.

x	$f^{-1}(x)$
-3	-2
0.5	-1
1	0
1.5	1

x	$f(x)$
-2	-3
-1	0.5
0	1
1	1.5

12. $f^{-1}(x)=\frac{1}{6}x+2$

13. $f^{-1}(x)=\sqrt[3]{\frac{1}{4}(x-3)}$

14. $f^{-1}(x)=x^2-3$

15. $f^{-1}(x)=\frac{x-2}{5}$

16. $f^{-1}(x)=\sqrt[3]{\frac{x}{3}}$

17. $f^{-1}(x)=(x-6)^2, x\geq 6$

18. $f^{-1}(x)=\frac{x-1}{3}$

19. $f^{-1}(x)=\sqrt[3]{x+4}$

20. $f^{-1}(x)=5x$

21. $x<-6, x>-3$ 22. $x<-2, x>\frac{1}{2}$

23. $-.5311\leq x\leq 7.5311$

24. $.6834\leq x\leq 7.3166$

25. $x\leq -.6180, x\geq 1.6180$

26. $x<-4.1926, x>1.1926$

27. $x<-1$ 28. $x<-1, x>3$

29. $x\leq 3, x\geq 7$ 30. $-12<x<4$

Section 4.4

1. x = 3 2. x = 22

3. x = 0, 2 4. x = -4

5. x = 2 6. x = 9

7. x = 3 8. x = 36

9. x = 0 10. x = 10.5

11. x = 7/4, -3/4 12. x = 8, 4

13. x = 1, 1/5 14. x = 3, -3

15. x = 0, -3 16. x = 6, -1

17. $-2<x<0$ 18. $x\leq 0, x\geq 5$

19. $-1\leq x\leq 4$ 20. $-3<x<5$

CHAPTER 5 ANSWER KEY

Section 5.1

1. a. not exponential b. exponential
c. not exponential d. exponential
e. exponential f. not exponential

2. a. growth b. decay
c. growth d. decay
e. decay f. growth

3. a.

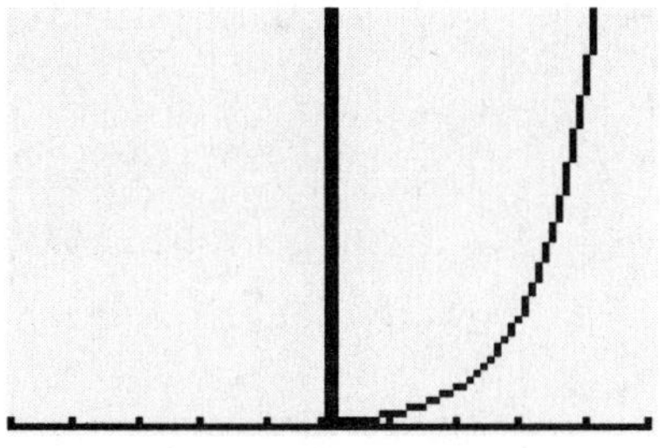

b. $f(1) = 3$, $f(-1) = 0.333$, $f(4) = 81$
c. $y = 0$ d. (0, 1)

4. a.

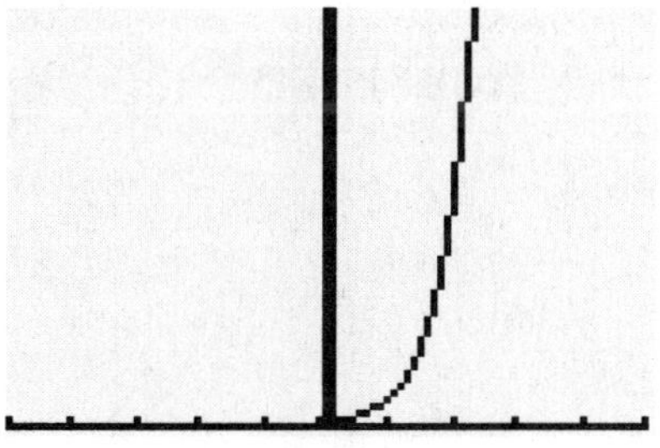

b. $f(1) = 7.389$, $f(-1) = 0.135$, $f(4) = 2980.958$
c. $y = 0$ d. (0, 1)

5.

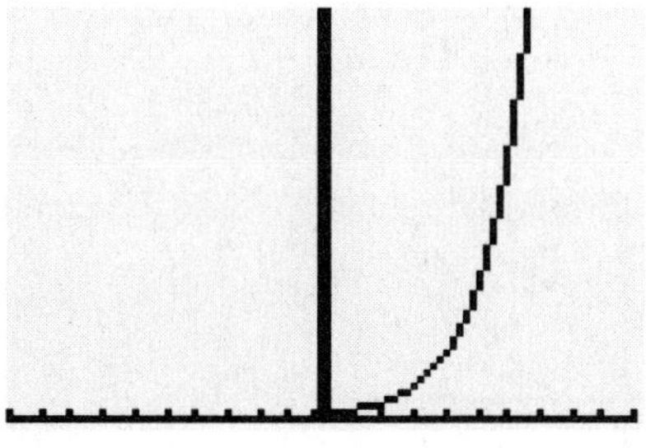

6.

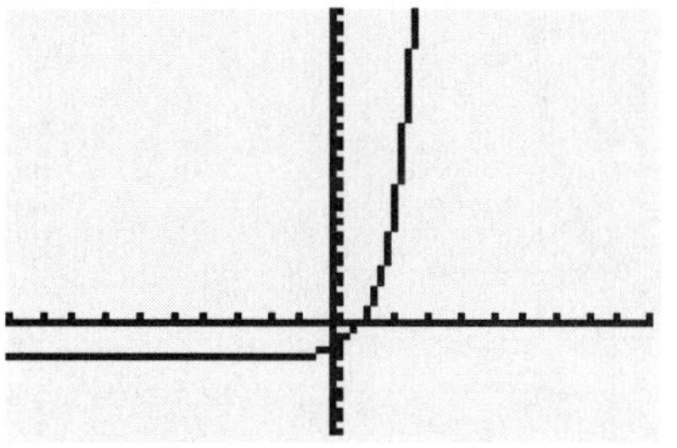

7.

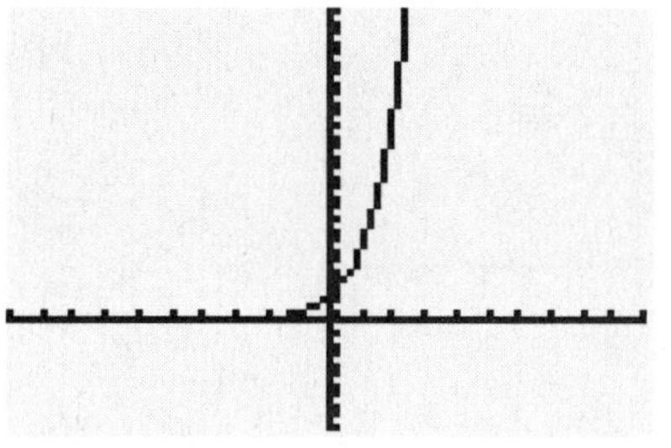

8.

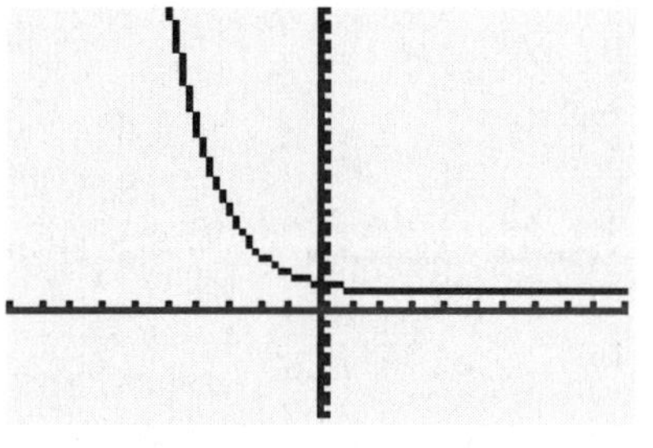

9.

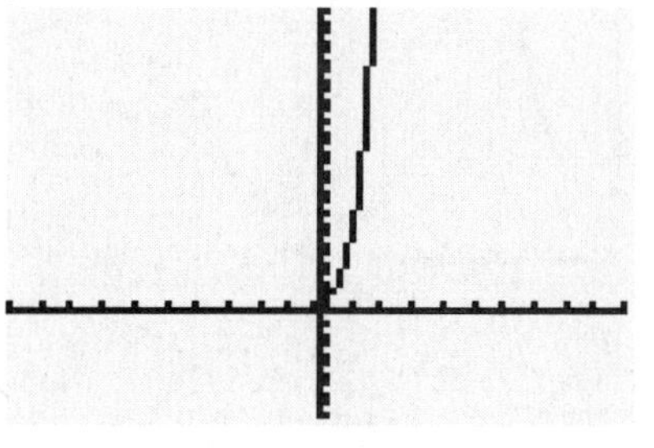

10.

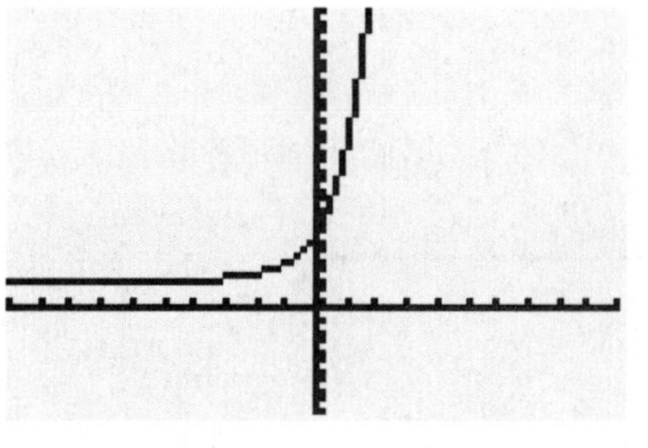

11.

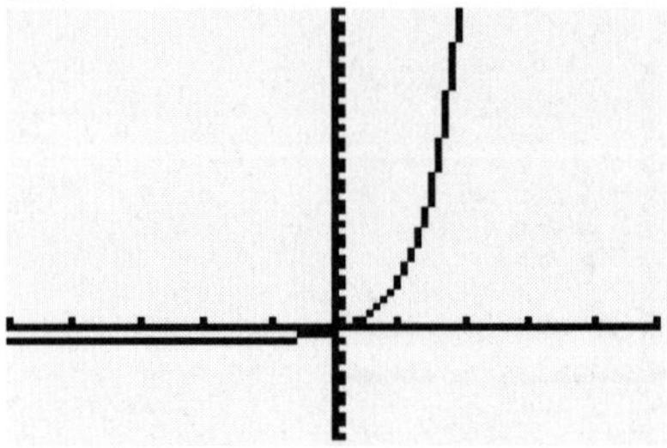

12.

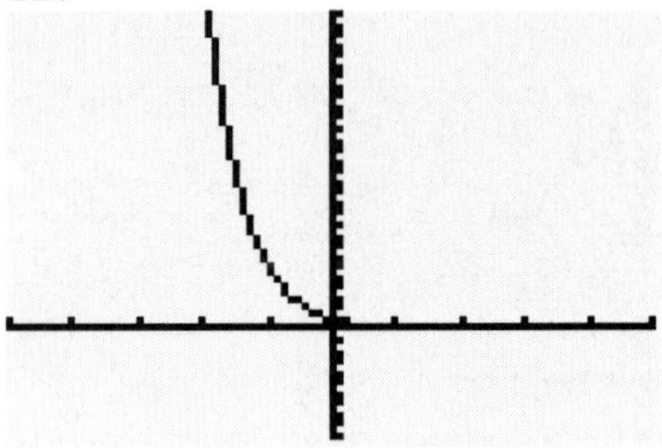

13.

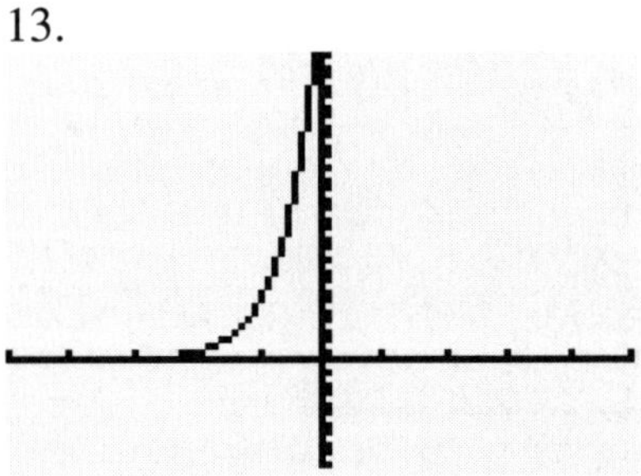

14.

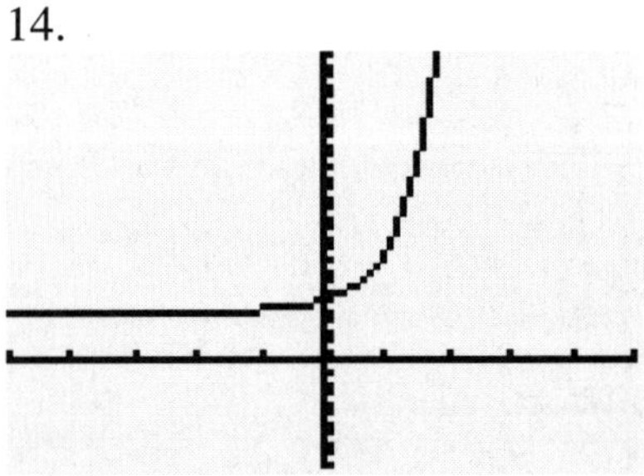

15.

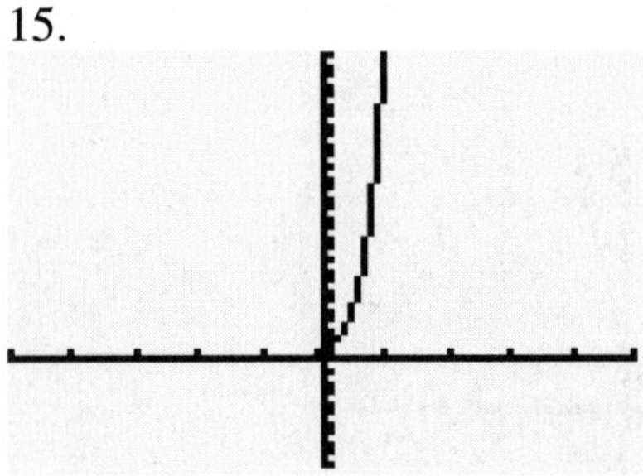

16.

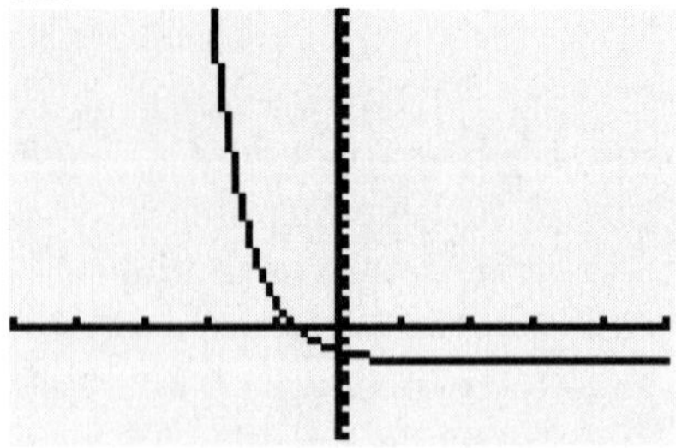

17. y = 7, (0, 8)

18. y = -1, (0, 3)

19. y = -2, (0, -1)

20. y = 0, (0, 0.0183)

Section 5.2

1. $2^y = x$

2. $10^y = 5$

3. $e^y = 3x$

4. $a^y = w$

5. $\log_3 x = b$

6. $\ln y = 3x$

7. $\log y = 3$

8. $\log_8 y = 3x$

9. 1.362

10. 11.145

11. not possible

12. 2

13. 3

14. 3

15. 3

16. -2

17. 0

18. 2

19. 4

20. 2

21. 1

22. ½

23. -2

24. 5/2

25.

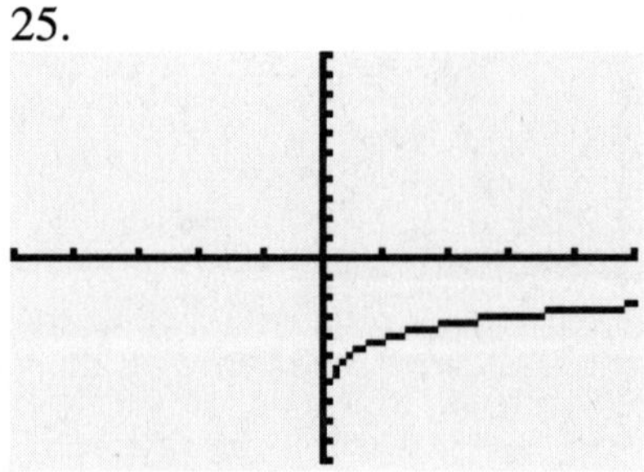

26.

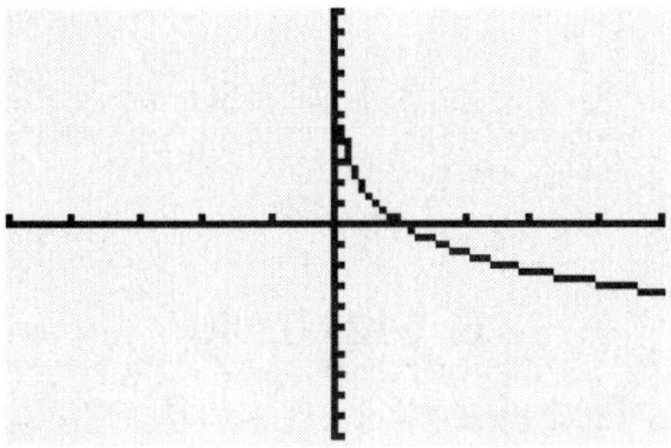

27.

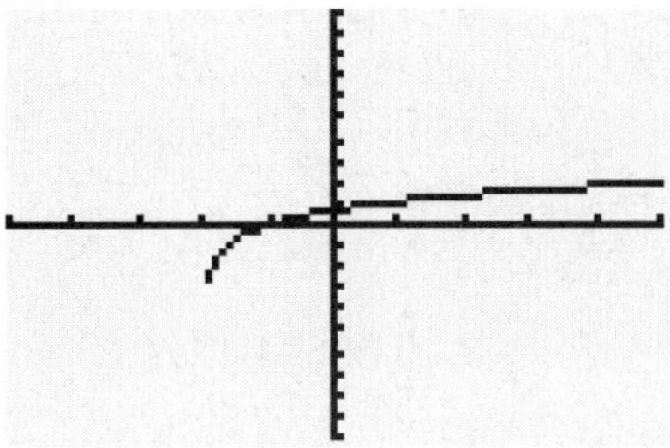

28.

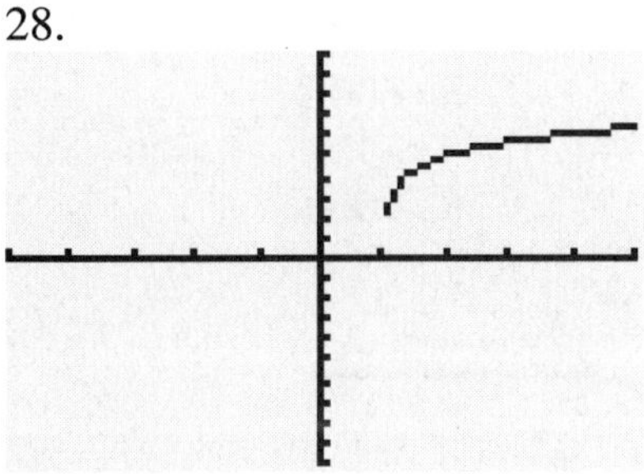

29. $\ln(3x+1)-\ln(2x-4)$

30. $4\log x+3\log(2x+1)$

31. $\frac{1}{5}\ln(4x+7)-(\ln 3+2\ln x)$

32. $\frac{1}{2}\left[\log(2x+1)-\log(x^2+5)\right]$

33. $\ln\left(\frac{x^6}{y^5}\right)$

34. $\log\sqrt[4]{x+6x^3}$

35. $\log\frac{(6x)^5y^2}{z}$

36. $\ln\sqrt[5]{xy^2}$

Section 5.3

1. x = 3.716
2. x = 1.185
3. x = 3.450
4. x = 1.596
5. x = 1.602
6. x = 3.689
7. 13
8. 7
9. 5
10. 2
11. 1.681
12. 0.477
13. 1.204
14. 0.540
15. 1.292
16. 7.229
17. 0.646
18. 0.290
19. $x=81$
20. $x=\frac{1}{2}$
21. $x=e^2$
22. $x=1000$
23. x = 7/4
24. x = 7/3
25. x = 2
26. x = 1.898
27. x = 9.453
28. x = 1.706
29. x = 2699.028
30. 162,752.79
31. x = 0.314
32. x = 5, 2
33. $x\le 3.746$
34. $x>5.524$

Section 5.4

1. $y=0.12(5)^{x+2}$

2. $f(x)$ is not exponential, $g(x)$ and $h(x)$ are exponential

3. $y=4.099(1.020)^x$

4. $y=123.56(1.621)^x$

5. a.

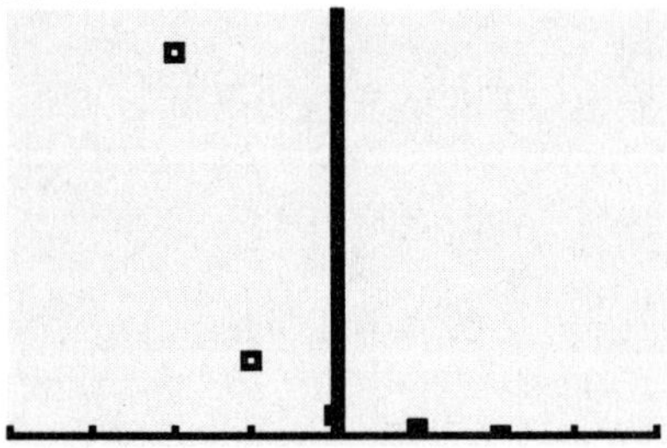

b. exponential

c. $y = 22.865(0.214)^x$

6. a.

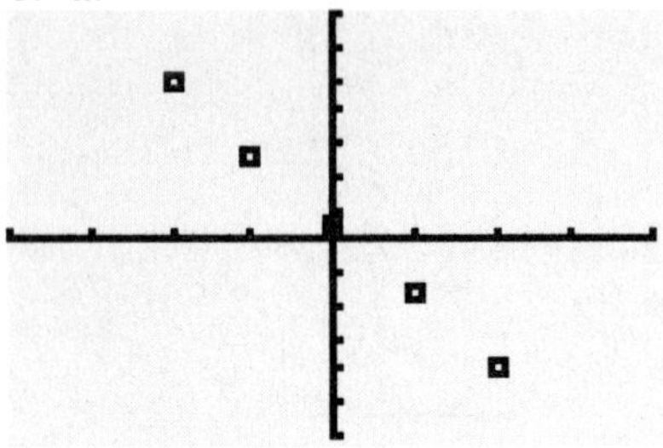

b. linear

c. $y = -2.2x + 0.5$

7. $y = 3.630 + 0.570 \ln x$

8. $y = 11.298 + 7.372 \ln x$

Section 5.5

1. 29,836.49 2. 9,316.64

3. 28,159.95 4. 503,132.84

5. 24,486.33 6. 989,602.82

7. 8,203.00 8. 48,754.39

9. 7,247,918.32 10. 175,319,949.80

11. $f(1) = 800$ $f(2) = 1000$
$f(3) = 1200$ $f(3.5) = 1200$
$f(4.5) = 1400$

12. $P = Ae^{-rn}$

13. \$3,885.72 14. \$9,033.45

Section 5.6

1. 181,638.83 2. 378,308.50

3. 3,306.211.13 4. 151.40

5. 108 6. \$12,603.40

7. \$7,243.28

8. a. \$2098.43 b. \$755,434.80
c. \$405,434.80

Section 5.7

1. 786 2. 0.00694

3. 100 4. $2.44(10)^{-40}$

5. 1593.34 6. 600

7. 1200 8. 1000

9.

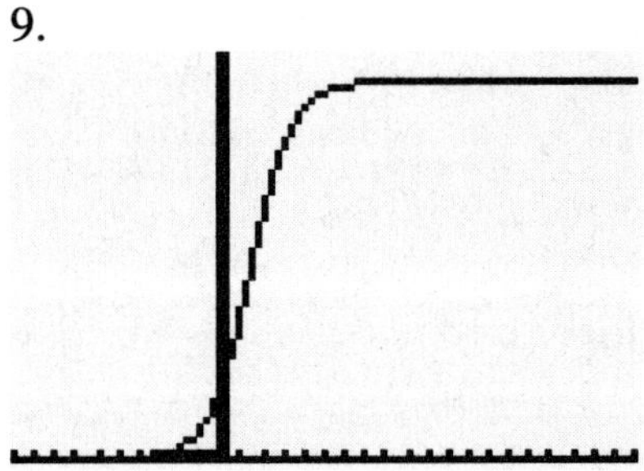

10. $f(0) = 140$, $f(10) = 699.87$

11. increasing 12. N = 700

13.

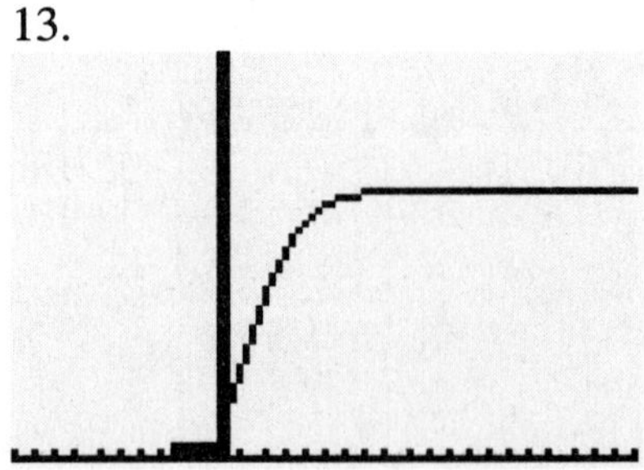

14. 24 15. N = 200

Section 6.1

1. c
2. d
3. b
4. d
5. a. (1, 0), (-1, 0), (0, 0)
   b. negative
   c. fourth degree
6. a. (0, 0), (-1, 0), (2, 0), (3, 0)
   b. positive
   c. fourth degree
7. a. 3rd degree; 3
   b. as $x \to +\infty, y \to +\infty$
   as $x \to -\infty, y \to -\infty$
   c.

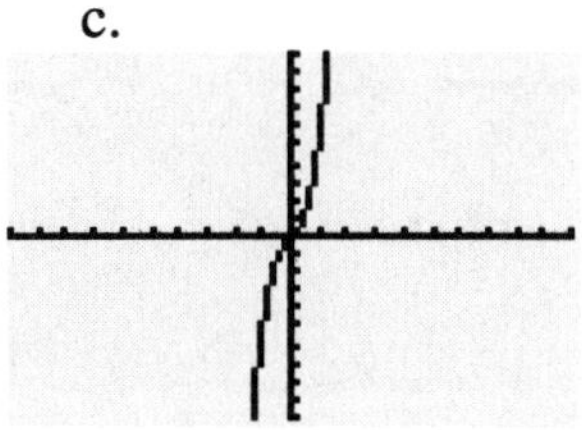

8. a. 4th degree; -0.01
   b. as $x \to \pm\infty, y \to -\infty$
   c.

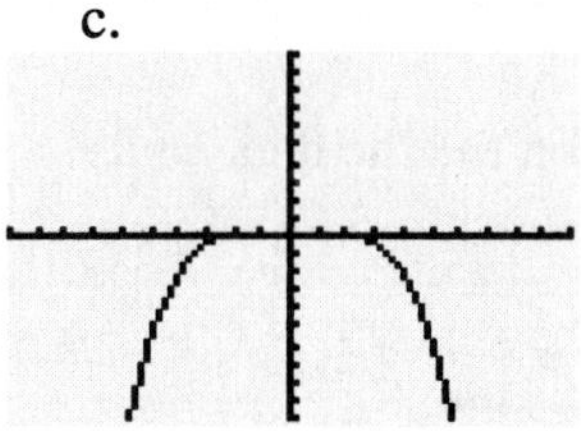

9. a. 3rd degree; -9
   b. as $x \to +\infty, y \to -\infty$
   as $x \to -\infty, y \to +\infty$
   c.

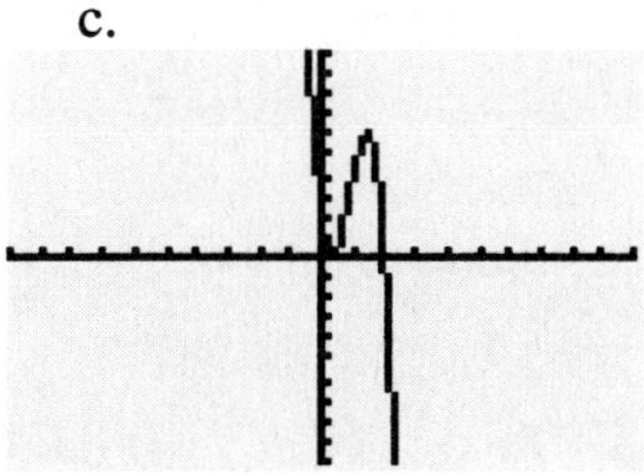

10. a. 3rd degree; 40
    b. as $x \to +\infty, y \to +\infty$
    as $x \to -\infty, y \to -\infty$
    c.

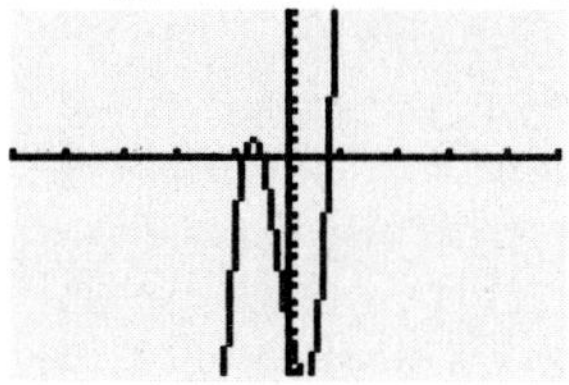

11. a. 3rd degree; 16
    b. as $x \to +\infty, y \to +\infty$
    as $x \to -\infty, y \to -\infty$
    c.

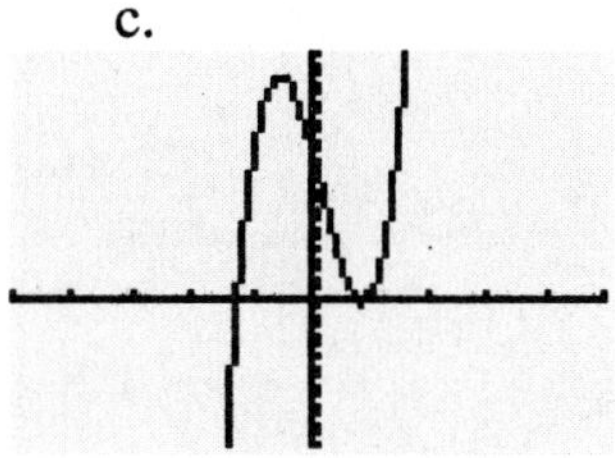

12. a. 4th degree; 8
    b. as $x \to \pm\infty, y \to +\infty$
    c.

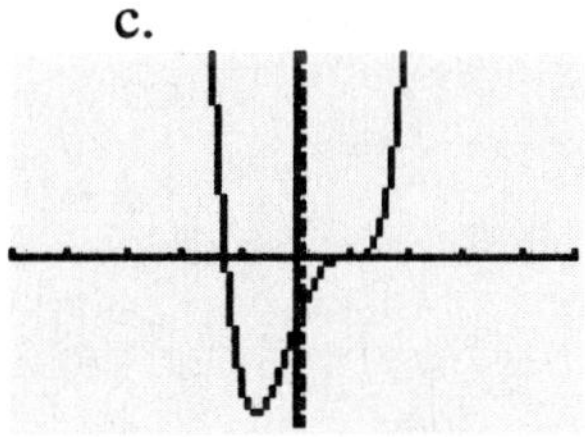

13.

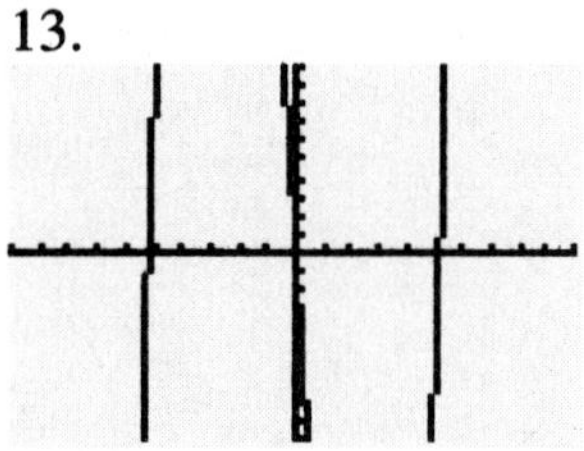

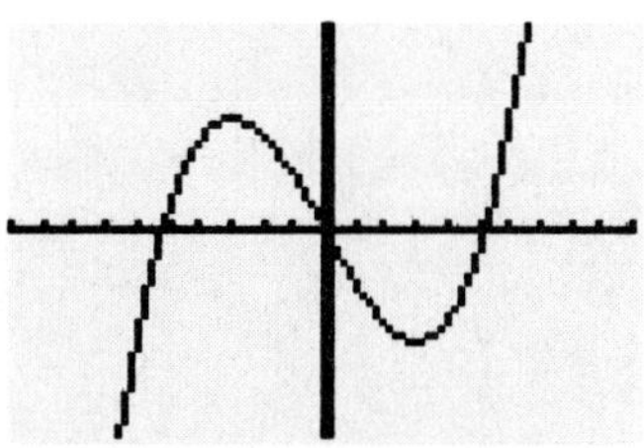

[-10, 10] by [-100, 100]

14.

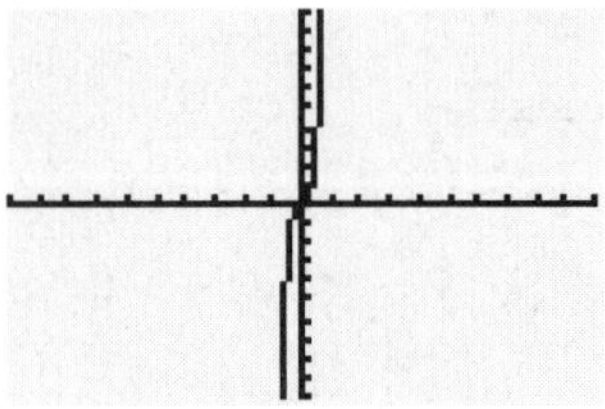

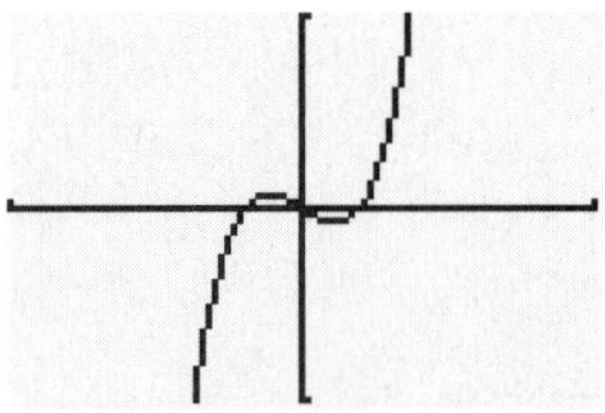

[-1, 1] by [-1,1]

15.

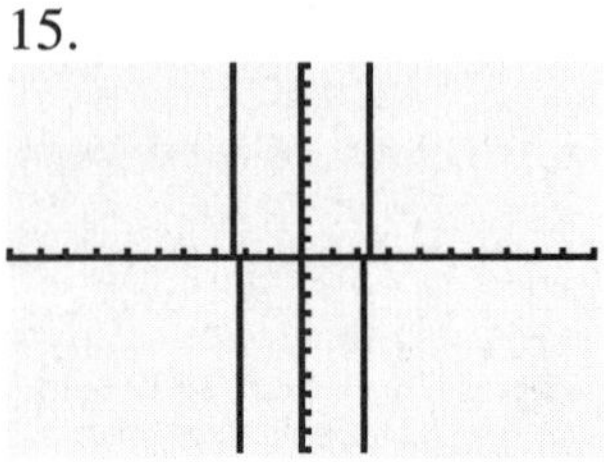

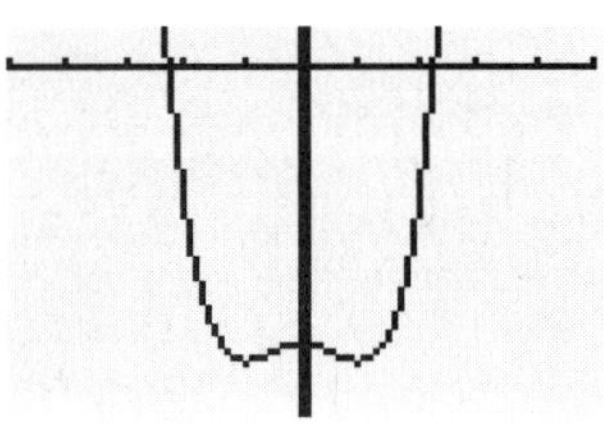

[-5, 5] by [-100, 10]

Section 6.2

1. $y = x^3 + 6x - 5$

2. $y = -10x^3 - 20x$

3. $y = 2x^4 + 5x^2 - 20$

4. $y = 0.1x^4 + 0.05x^2$

5. a.

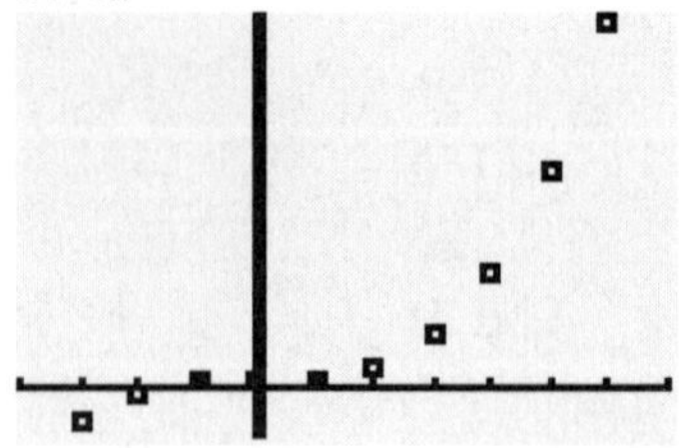

b. cubic

6. a. $y = 0.488x^3 + 0.24x^2 + 0.131x + 1.559$
b. $y = 10.644x + 5.804$
c. The cubic function fits the data better.

7. a.

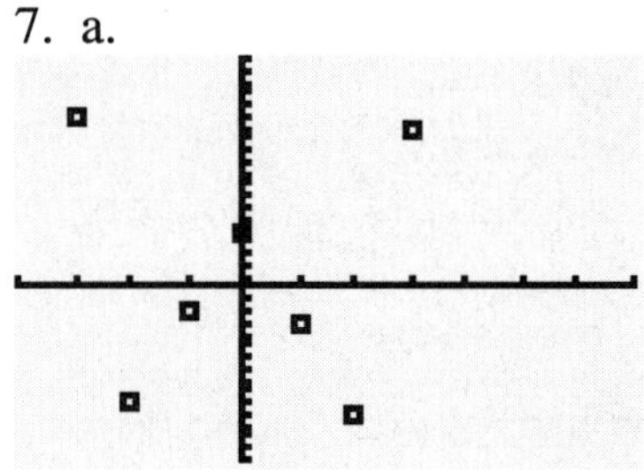

b. quartic

8. a. $y = 0.28x^3 + 1.667x^2 - 0.409x - 5.524$
b. $y = 0.981x^4 + 0.028x^3 - 7.723x^2$
$-0.409x + 4.567$
c. The quartic function fits the data better.

9. a.

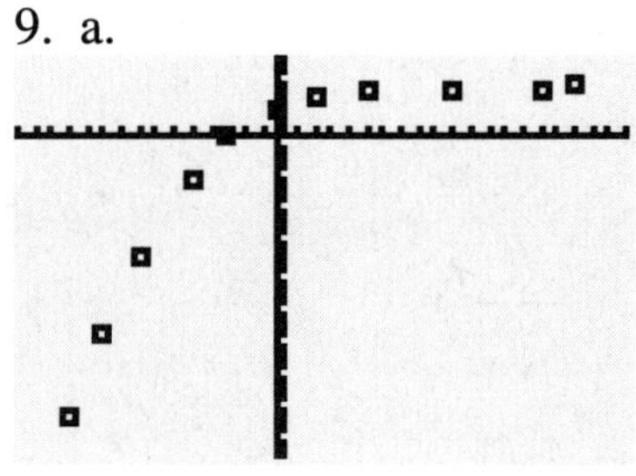

b. cubic

10. a.
$y = 0.005x^3 - 0.131x^2 + 1.003x + 3.917$
b. $y = 1.19x - 5.366$
c. The cubic function fits the data better.

11. a.

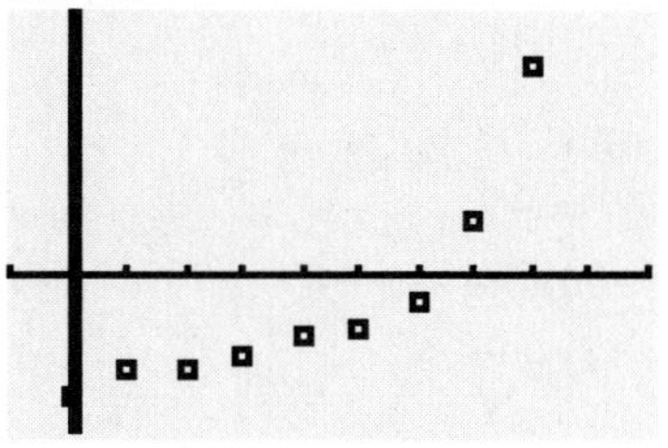

b. cubic

12. a.
$y = 0.402x^3 - 3.417x^2 + 9.409x - 24.23$
b.
$y = 0.07x^4 - 0.68x^3 + 1.96x^2 + 0.99x - 22.6$
c. The cubic and the quartic are essentially the same for the data.

Section 6.3

1. $x = 0, 2, -2$
2. $x = 0, 4, -4$
3. $x = 0, 6$
4. $x = 0, 6, -5$
5. $x = -3, -1, 2$
6. $x = 0, 8$
7. $x = 0, 1, 1$
8. $x = 0, 5, 3$
9. $x = 0, -6, -1$
10. $x = 0, -2, 3$
11. $x = -3, -2, 2$
12. $x = -3, -2, 3$
13. $x = -1, 1, 7$
14. $x = -5, -1/2, 5$
15. $x = 2$
16. $x = 4$
17. $x = 3$
18. $x = 2$
19. x = 0, ½, -1/2
20. $x = 0, \sqrt{\frac{10}{3}}, -\sqrt{\frac{10}{3}}$
21. $x = 0, \sqrt{\frac{3}{2}}, -\sqrt{\frac{3}{2}}$
22. $x = 0, \sqrt{56}, -\sqrt{56}$
23. $x = 0, \sqrt{30}, -\sqrt{30}$
24. $x = -2, 2$
25. $x = \pm\sqrt{10}$
26. $x = -0.5, 0.5$
27. $x = -0.758, 2.099, 5.659$
28. $x = -2.946, 0.252, 2.694$

Section 6.4

1. $x^3 - 5x^2 + 12x - 20 + \frac{58}{x+2}$
2. $3x^3 + 7x^2 + 21x + 3 + \frac{4}{x-3}$
3. $x^4 - x^3 + 4x^2 + 2x - 2 - \frac{3}{x+1}$
4. $x^4 - 2x^3 + 4x^2 - 5x + 10 - \frac{18}{x+2}$
5. yes
6. no
7. no
8. yes
9. $x = -3, 5$
10. $x = -3, -4$
11. $x = -4, 6$
12. $x = -2, -1$
13. $x = -2, -3, \frac{1}{2}$

14. $x = -3, 0, 1$

15. $x = -4, 4, -\frac{1}{4}$

16. $x = -2, 6, 7$

17. $\pm1, \pm2, \pm3, \pm4, \pm6, \pm12$

18. $\pm1, \pm2, \pm5, \pm10$
$\pm\frac{1}{4}, \pm\frac{1}{2}, \pm\frac{5}{4}, \pm\frac{5}{2}$

19. $x = 2, 3, 4$

20. $x = -3, -2, 2$

21. $x = -1, -1, 3$

22. $x = -4, 1, 2, 2$

23. $x = 5, \pm 2i$

24. $x = -1, \pm\sqrt{5}i$

25. $x = -3, \frac{1 \pm \sqrt{7}i}{2}$

26. $x = 4, \pm 3i$

27. $x = -2, \frac{-3 \pm \sqrt{3}i}{2}$

28. $x = \frac{3}{2}, -1 \pm \sqrt{3}i$

Section 6.5

1. a. $x = -2$ b. $y = 0$

2. a. $x = -1$ b. $y = 0$

3. a. $x = -\frac{5}{2}$ b. $y = \frac{1}{2}$

4. a. $x = 4$ b. $y = -2$

5. a. no vertical asymptote
   b. no horizontal asymptote

6. a. no vertical asymptote
   b. $y = 1$

7. c

8. c

9. a. $x = -2$ b. $y = 1$
   c.

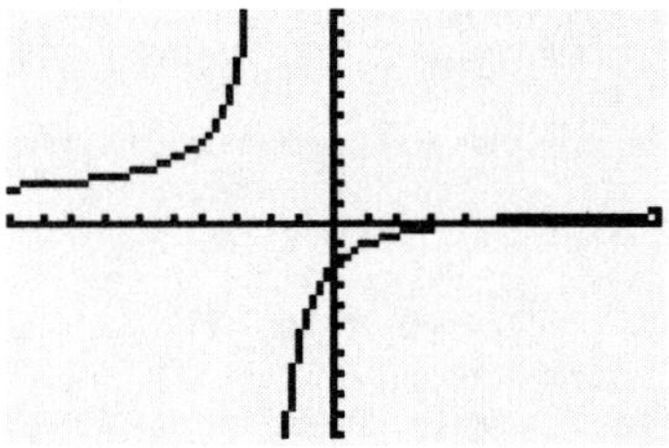

10. a. $x = 6$ b. $y = 3$
    c.

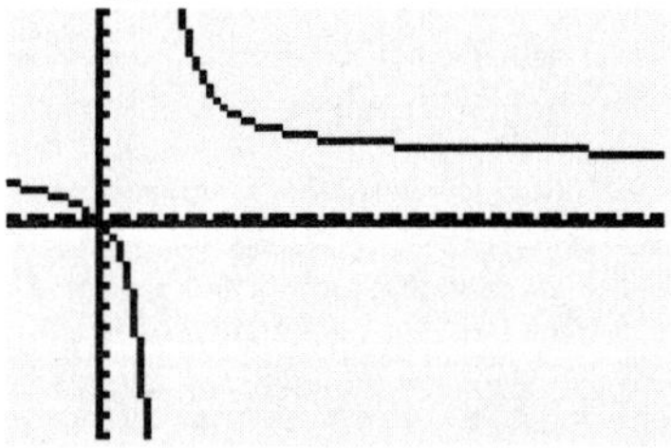

11. a. $x = 3, -3$ b. $y = 0$
    c.

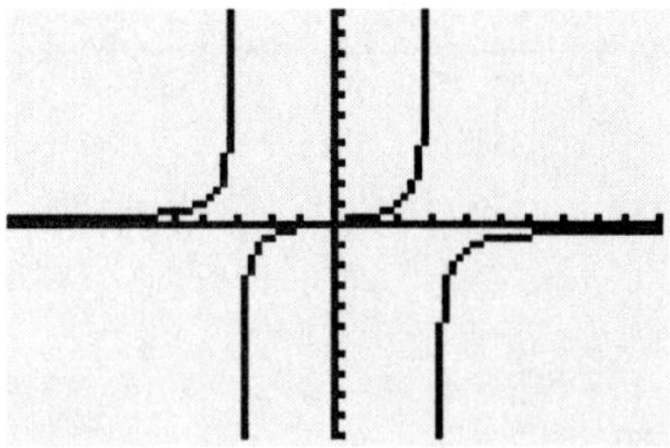

12. a. $x = 1$
    b. no horizontal asymptote
    c.

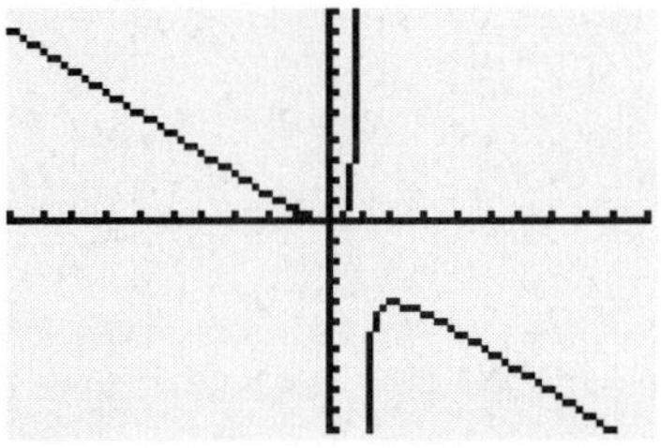

13. a.

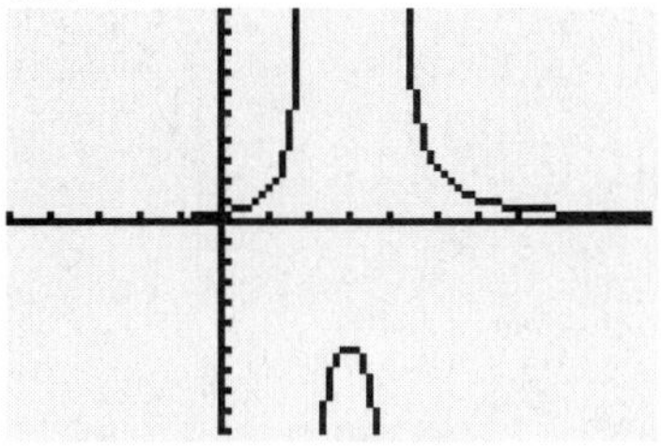

b. When $x = -2$, $y = \frac{1}{24}$

When $x = 0$, $y = \frac{3}{8}$

c. $x = -3$ d. $x = 3$

14. a.

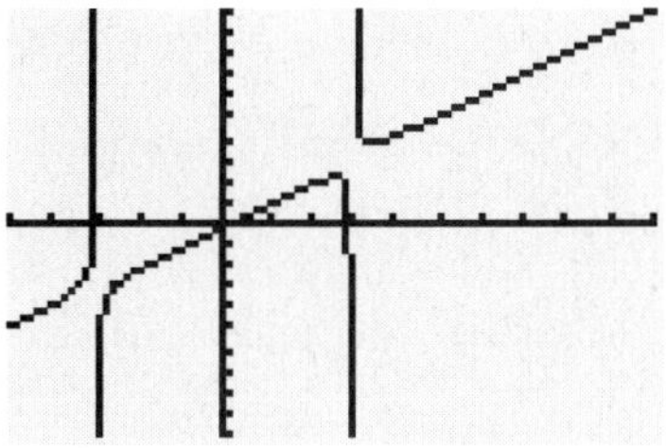

b. When $x = -2$, $y = \frac{1}{5}$

When $x = 2$, $y = -\frac{3}{5}$

c. $x = -5, -1$ d. $x = 2, -3\frac{2}{3}$

15. a.

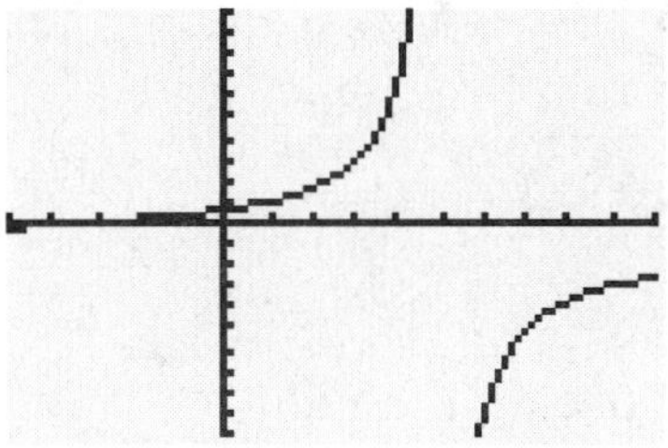

b. When $x = -2$, $y = \frac{1}{7}$

When $x = 1$, $y = 1$

c. $x = 4$ d. $x = 3$

16. $x = -1.5$

17. $x = -\frac{1}{4}$

Section 6.6

1. $x \le 0, x \ge \frac{1}{3}$

2. $-\frac{1}{2} \le x \le \frac{1}{2}$

3. $x \le -3, 0 \le x \le 3$

4. $x < -3, x > 1$

5. $x \le 4$

6. $x < \frac{1}{2}, x > 3$

7. $-4 < x < 0$

8. $0 < x < 1$

9. $x < -0.828, x > 4.828$

10. $x < 0$

11. $x \le 2, x \ge 8$

12. $x > -2$

13. a. $x < 2$
    b. $x \geq 2$

14. a. $x < -1, 0 < x < 2$
    b. $-1 \leq x \leq 0$

15. a. $-1 < x < 0, 0 < x < 1$
    b. $x \leq -1, x \geq 1, x = 1$

Section 7.1

1. x = 2, y = 6, z = 4 2. x = -1, y = 4, z = -3

3. x = 7, y = 0, z = 1 4. x = 3, y = -1, z = 1

5. x = 2, y = -3, z = 4 6. x = 16, y = 8, z = -2

7. x = 4, y = 3, z = -1 8. x = 6, y = -2, z = 1

9. x = 5, y = -1, z = 6 10. x = -1, y = 0, z = 2

11. x = 1, y = -4, z = 2 12. x = -1, y = ½, z = 2

13. x = -3/2, y = 1, z = -5

14. x = 5, y = -4, z = 2

15. x = -3z – 2, y = z + 4, z = z (any number)

16. x = 7z - 1, y = -3z + 2, z = z

17. x = -z + 2, y = -3z + 2, z = z

18. no solution 19. no solution

20. x = 7z, y = 2z – 2, z = z

Section 7.2

1. $\left[\begin{array}{ccc|c} 1 & 2 & -1 & 4 \\ 2 & 1 & 4 & 8 \\ -1 & 3 & -2 & 6 \end{array}\right]$ 2. $\left[\begin{array}{ccc|c} 3 & 2 & 1 & 8 \\ -2 & 3 & 4 & 2 \\ 1 & 4 & -1 & 6 \end{array}\right]$

3. $\left[\begin{array}{ccc|c} 2 & 4 & 5 & 11 \\ 3 & -1 & 2 & 4 \\ 5 & 1 & -1 & 5 \end{array}\right]$ 4. $\left[\begin{array}{ccc|c} 3 & 2 & 2 & 7 \\ 5 & -2 & 4 & 7 \\ 8 & -2 & -1 & 5 \end{array}\right]$

5. x = 3, y = -2, z = 4

6. x = -2, y = 6, z = 9

7. x = -3, y = 2, z = -1

8. x = -1, y = 1, z = -3

9. x = 4, y = 1, z =1

10. no solution

11. x = -3z + 2, y = -6z + 5, z = z

12. x = 5, y = -6, z = -2

13. no solution

14. x = 2, y = 0, z = -1

15. x = ½, y = -3, z = 4

16. x = -3, y = 4, z = 2

17. x = 2, y = 3, z = 1, w = 1

18. x = 1, y = 0, z = 2, w = -1

19. x = -1, y = -4, z = 2

20. x = -2z + 4, y = -3z + 1, z = z

21. no solution

22. x = 3/2, y = -2/3, z = -1/4

Section 7.3

1. x = 3, y = 2, z = 3, w = -1

2. $\begin{bmatrix} 11 & 4 \\ 0 & 10 \end{bmatrix}$ 3. $\begin{bmatrix} -9 & -6 \\ 5 & -5 \end{bmatrix}$

4. no solution

5. $\begin{bmatrix} -8 & 11 \\ 2 & 16 \\ -17 & 22 \end{bmatrix}$ 6. no solution

7. $\begin{bmatrix} -6 & 9 & 12 \\ 0 & 18 & 16 \end{bmatrix}$ 8. $\begin{bmatrix} -11 & 20 \\ 19 & -16 \end{bmatrix}$

9. no solution

10. $\begin{bmatrix} 4 & -42 & 24 \\ 38 & -84 & -52 \\ 15 & -63 & 6 \end{bmatrix}$

11. $\begin{bmatrix} 36 & 18 \\ 40 & 76 \end{bmatrix}$

12. $\begin{bmatrix} 25 & 0 \\ -12 & 1 \end{bmatrix}$

13. $\begin{bmatrix} 9 & 16 \\ 8 & 17 \end{bmatrix}$

14. $\begin{bmatrix} 5 & 20 \\ 0 & -5 \end{bmatrix}$

15. $\begin{bmatrix} -3 & 4 \\ 4 & 3 \end{bmatrix}$

16. $\begin{bmatrix} 8 & 7 & 18 \\ 3 & 8 & 7 \\ 7 & -21 & 5 \end{bmatrix}$

17. $\begin{bmatrix} -2 & -5 & 16 \\ -5 & -14 & 10 \\ -2 & -8 & -17 \end{bmatrix}$

18. $\begin{bmatrix} 13 & -10 & 5 \\ 16 & -23 & 4 \\ -2 & -12 & 0 \end{bmatrix}$

19. $\begin{bmatrix} -2 & -7 & 15 \\ 2 & 7 & 12 \\ 2 & -2 & -15 \end{bmatrix}$

20. \$35.60 for 20 lbs. of Morning Glory, \$30.70 for Wake Up Call

Section 7.4

1. a. $AB = \begin{bmatrix} 1 & 0 \\ 0 & 1 \end{bmatrix}$ b. $BA = \begin{bmatrix} 1 & 0 \\ 0 & 1 \end{bmatrix}$
c. yes, $AB = BA = I$

2. a. $AB = \begin{bmatrix} 1 & 0 \\ 7 & -13 \end{bmatrix}$ b. $BA = \begin{bmatrix} 1 & 21 \\ 0 & -13 \end{bmatrix}$
c. no, $AB \neq BA \neq I$

3. $\begin{bmatrix} \frac{1}{6} & \frac{2}{3} \\ \frac{1}{6} & \frac{-1}{3} \end{bmatrix}$

4. $\begin{bmatrix} \frac{1}{14} & \frac{-4}{14} \\ \frac{3}{14} & \frac{2}{14} \end{bmatrix}$

5. $\begin{bmatrix} \frac{5}{13} & \frac{6}{13} \\ \frac{-3}{13} & \frac{-1}{13} \end{bmatrix}$

6. does not exist

7. $\begin{bmatrix} 1 & -1 \\ -3.5 & 4 \end{bmatrix}$

8. $\begin{bmatrix} 0 & 0 & 1/3 \\ -1/2 & -1/2 & 1/2 \\ 1 & 0 & -1/3 \end{bmatrix}$

9. $\begin{bmatrix} 2 & 1 & -1 \\ 8 & 2 & -5 \\ -11 & -3 & 7 \end{bmatrix}$

10. $\begin{bmatrix} \frac{7}{4} & \frac{5}{2} & 3 \\ \frac{-1}{4} & \frac{-1}{2} & 0 \\ \frac{-1}{4} & \frac{-1}{2} & -1 \end{bmatrix}$

11. $X = \begin{bmatrix} 1 \\ 6 \end{bmatrix}$

12. $X = \begin{bmatrix} 0 \\ 9 \\ 4 \end{bmatrix}$

13. x = 8, y = 6

14. x = 11, y = -3, z = 5

15. x = 1, y = 0, z = -1

16. x = -3, y = 4, z = 0

17. x = -31, y = 24, z = -4

18. x = -1, y = 1, z = -3, w = -2

19. x = -2, y = 1, z = 0, w = -1

Section 7.5

1. $x = 1, y = -1$ and $x = 3, y = 7$

2. $x = 1, y = 4$ and $x = 2, y = 5.5$

3. $x = -1, y = 3$ and $x = 6, y = -4$

4. $x = 1, y = 2$ and $x = -1, y = -2$

5. $x = 5, y = 2$ and $x = 2, y = 5$

6. $x = 3, y = 1$ and $x = -1, y = -7$

7. $x = 5, y = 2$ and $x = -7, y = -22$

8. $x = 4, y = 10$ and $x = 2, y = 2$

9. $x = 4, y = 1$ and $x = -5, y = -26$

10. $x = -1, y = 3$ and $x = -5, y = -1$

11. $x = 2, y = 4$ and $x = -2, y = 4$

12. $x = 4, y = 3$ and $x = \frac{1}{2}, y = 10$

13. $x = 0, y = 3$ and $x = 2, y = 1$
and $x = -2, y = 5$

14. $x = 1, y = -2$, and $x = -1, y = -10$
and $x = 7, y = 70$

15. $x = 0, y = 8$ and $x = 1, y = 6$

Section 8.1

1.

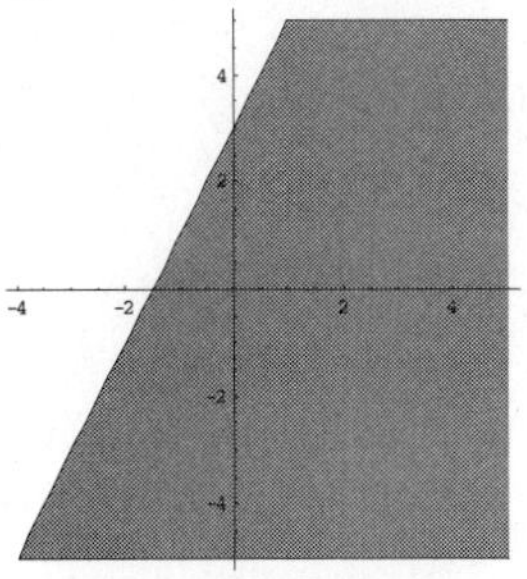

2.

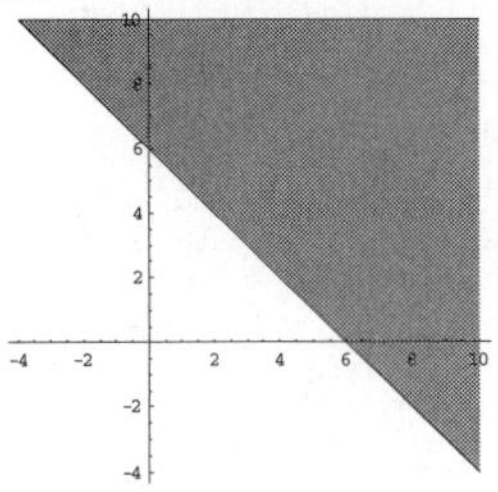

3.

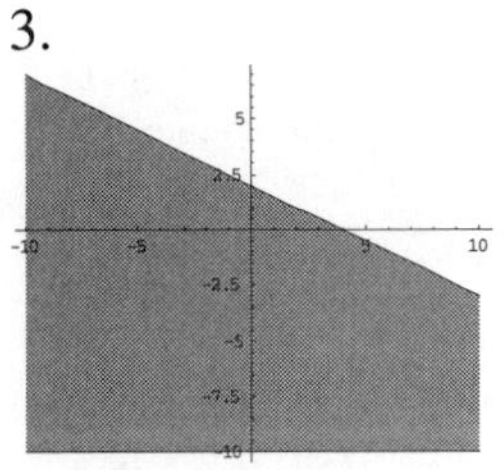

4.

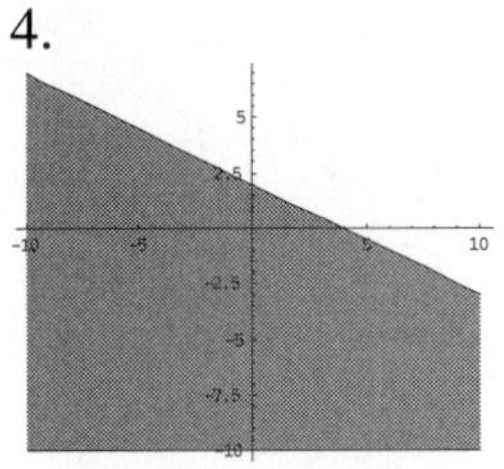

5.

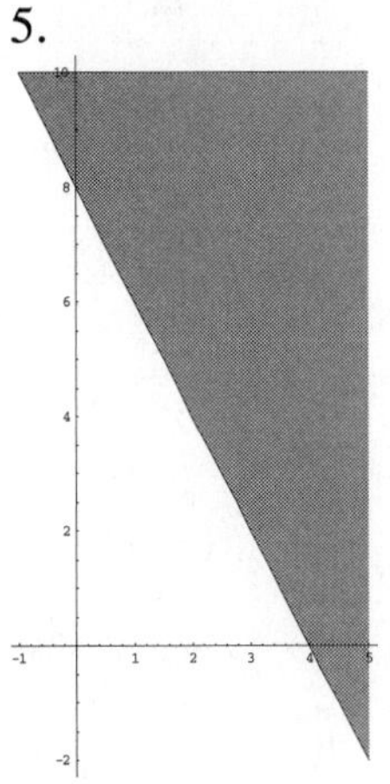

6.

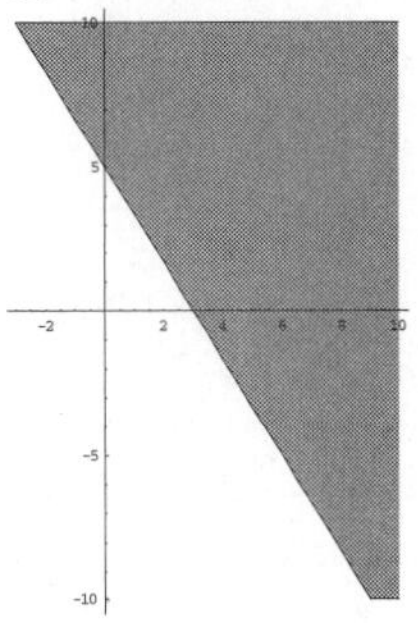

7.

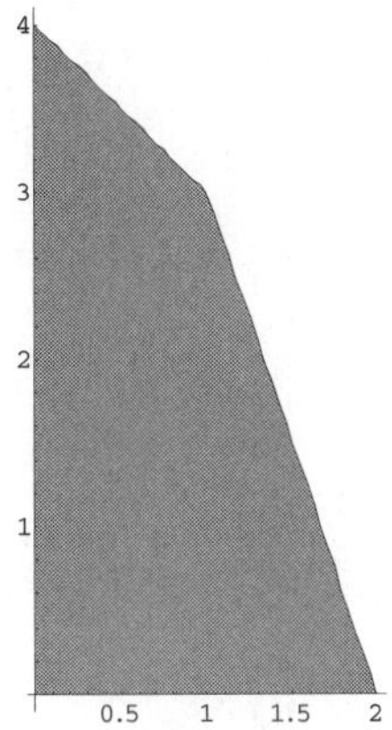

8. (0, 0), (0, 4), (1, 3), (2, 0)

9.

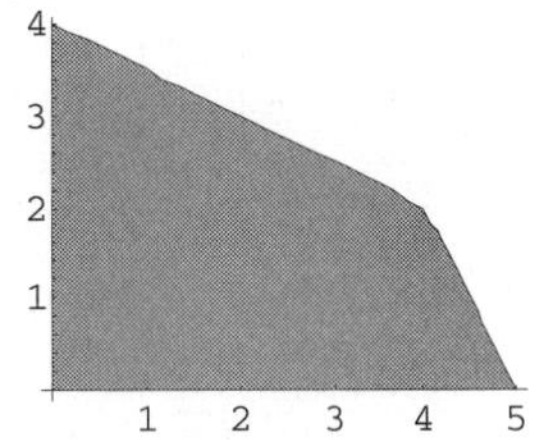

10. (0, 0), (0, 4), (4, 2), (5, 0)

11.

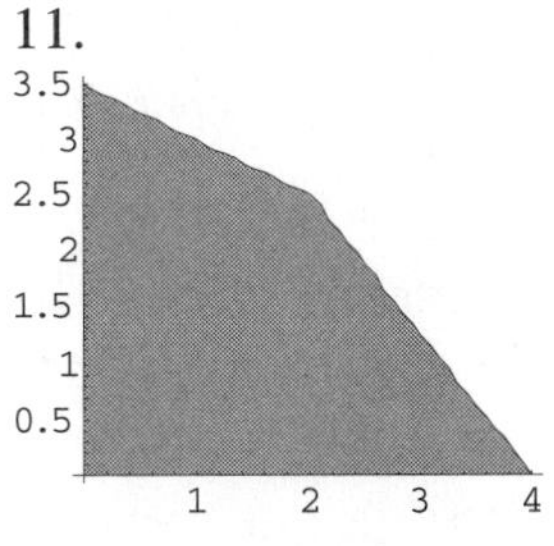

12. (0, 5), (7, 0), (2, 2.5)

13.

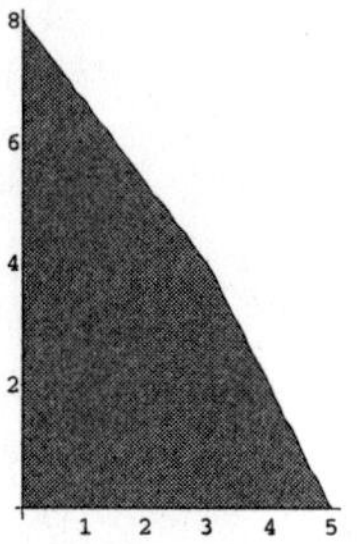

14. (0, 0), (0, 8), (3, 4), (5, 0)

15.

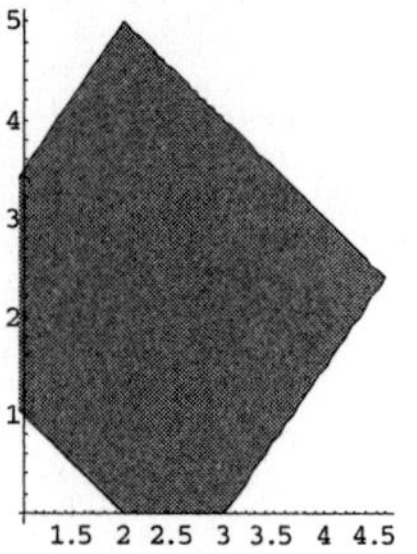

16. ((0, 2), (2, 5), (3, 0), (4, 3), (2, 0)

17.

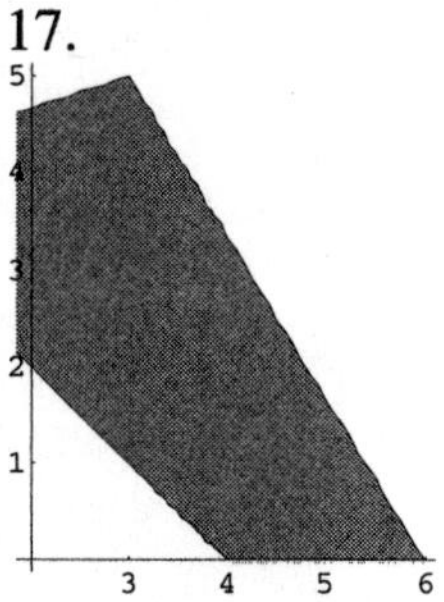

18. (0, 4), (3, 5), (4, 0), (6, 0)

19.

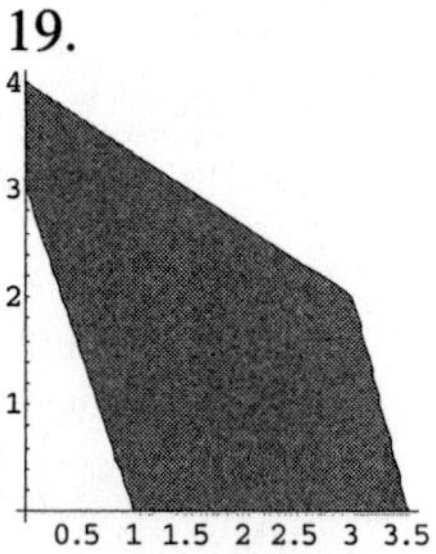

20. (0, 3), (0, 4), (3, 2), (3.5, 0)

21.

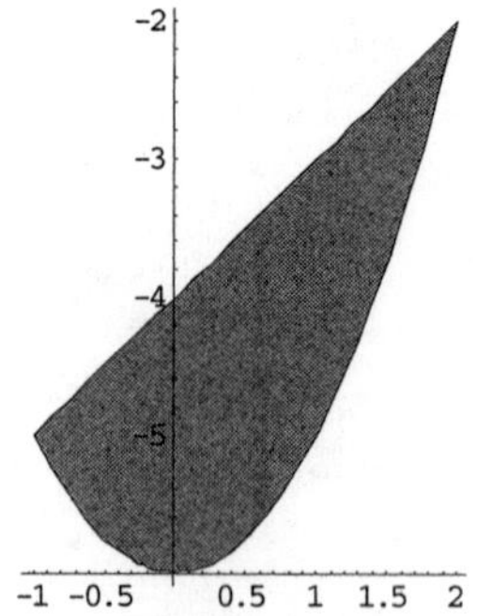

22.

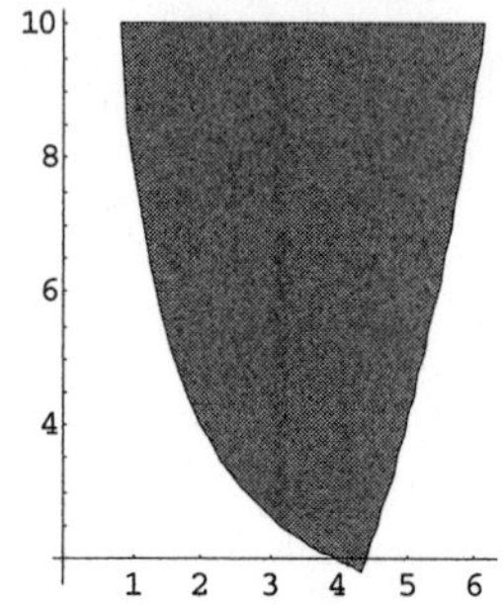

23.

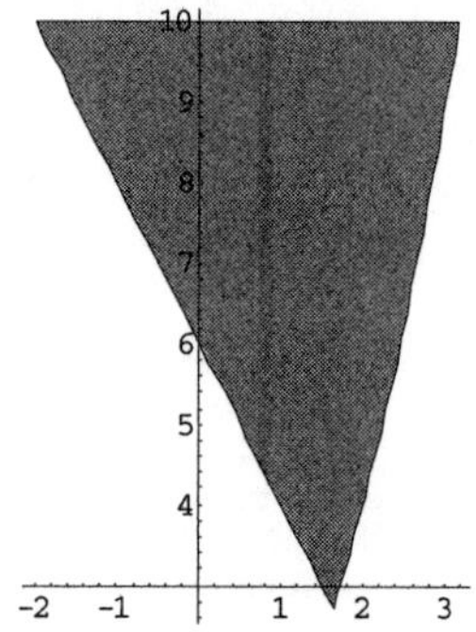

24.

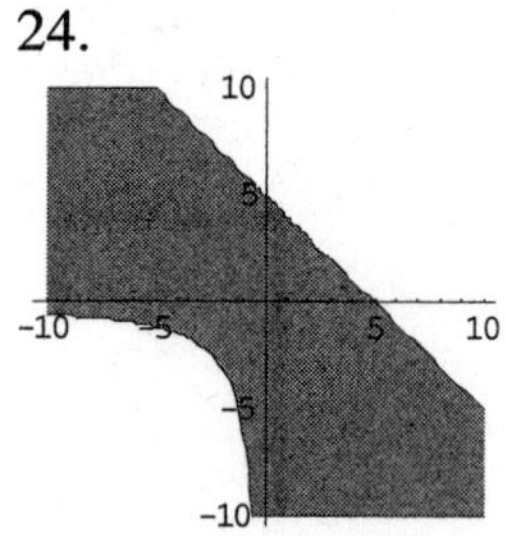

Section 8.2

1.

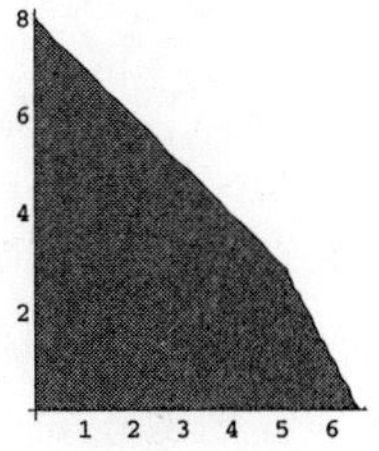

2. (0, 0), (0, 8), (5, 3), (6.5, 0)

3. Maximum of 29 at (5, 3)

4. Minimum of 0 at (0, 0)

5.

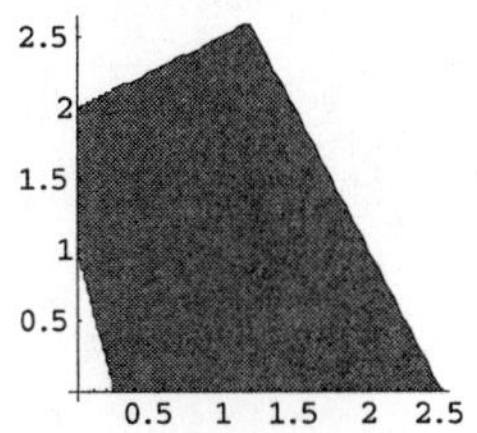

6. (0, 1), (0, 2), (0.25, 0), (1.2, 2.6), (2.5, 0)

7. Maximum of 3.8 at (1.2, 2.6)

8. Minimum of 0.25 at (0.25, 0)

9.

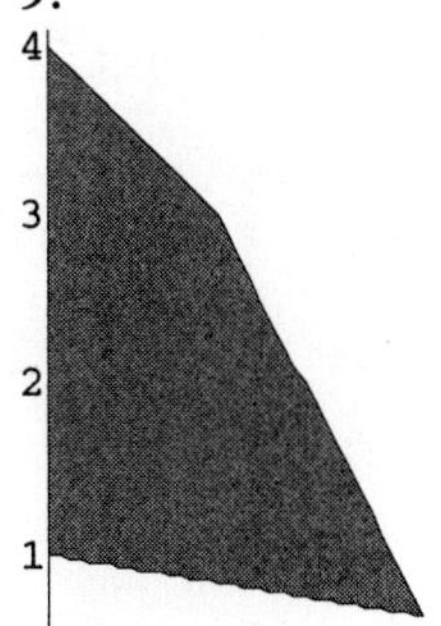

10. (0, 1), (0, 4), (2, 2/3), (1, 3)

11. Maximum of 20 at (0, 4)

12. Minimum of 5 at (0, 5)

13.

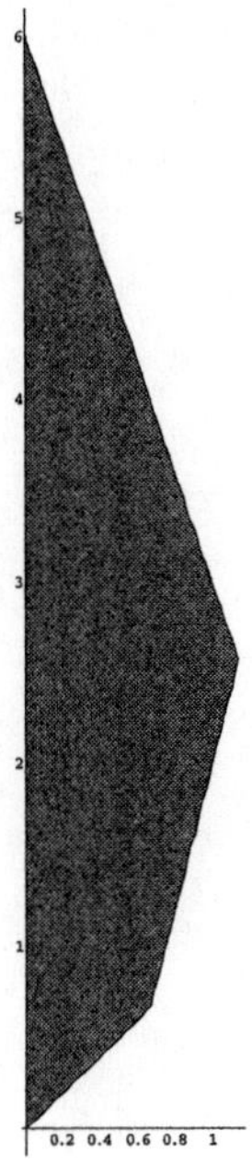

14. (1.14,2.57), (1.5, 1.5), (0, 0), (0, 6)

15. Maximum of 24 at (0, 6)

16. Minimum of 0 at (0, 0)

17. Maximum of 6 at (1.5, 1.5)

18. Minimum of 0 at (0, 0)

Section 8.3

1. 5, 8, 11, 14, 17, . . .

2. -4, 9, -16, 25, -36, . . .

3. 1 ½, 2 1/3, 3 ¼, 4 1/5, 5 1/6, . . .

4. 60, 30, 20, 15, 12, . . .

5. 16, 19, 22

6. -14, -17, -20

7. 43

8. 109.5

9. 81/2, 243/2, 729/2

10. 126.56, 94.92, 71.19

11. .005

12. 73,715,548.81

13. 6, 10, 14, 18

14. 10, 8, 6, 4

15. 8, 10, 11, 11.5, 11.75

Section 8.4

1. 128
2. 31.97
3. -820
4. 1207.5
5. 1425
6. -3300
7. 3,486,784,400
8. 21.33
9. 132,970.27
10. -3.33
11. 62.5
12. Sum does not exist
13. 9840
14. 329,554,456
15. 0.333
16. 1.74
17. -0.14
18. 9972.77
19. 1.5
20. 3

Section 8.5

1. $(3x+5)^3(2x-9)^4(23x-11)$

2. $(24x+24)(3x^2+6x-9)^3$

3. $(x^2+6)^4(3x-8)^2(39x^2-80x+54)$

4. $\dfrac{(x+3)(-x-1)}{(x-5)^2}$

5. $f'(3)=65$

6. $f'(-2)=3$

7. $f'(1)=\dfrac{-85}{64}$

8. $f'(2)=6e^{10}$

9. a. $6(x+h)-5$
   b. $6h$
   c. 6

10. a. $4-2(x+h)$
    b. $-2h$
    c. -2

11. a. $\dfrac{x+h}{2}+4$
    b. $\dfrac{h}{2}$
    c. $\dfrac{1}{2}$

12. a. $\dfrac{2-(x+h)}{3}$
    b. $-\dfrac{h}{3}$
    c. $-\dfrac{1}{3}$

13. $y=\dfrac{5}{8}x^3-4$

14. $y=\dfrac{9}{2}x^2+3$

15. $y=\dfrac{3}{4}x^2+\dfrac{3}{2}x-\dfrac{1}{2}$

16. $y=-\dfrac{3}{4}x^2+2x+3$

17. a. $x=3$ b. $(3,-7)$

18. a. $x=5$ b. $(5,150)$

19. a. $x = -2$ b. $(-2, -17)$

20. a. $x = 0, -4$
  b. $y = 2$ when $x = 0$
  $y = 34$ when $x = -4$
  c. Maximum of 34 when $x = -4$

21. a. $x = 1, -1$
  b. $y = 9$ when $x = 1$
  $y = 7$ when $x = -1$
  c. Maximum of 9 when $x = 1$

22. $y = 5x^{1/2}$

23. $y = 8x^{1/5}$

24. $y = 7x^{4/3}$

25. $y = 8x^{5/2}$

26. $y = (4x + 7)^{1/2}$

27. $y = (2x + 3)^{3/2}$

28. $y = (x^2 - 6)^{1/3}$

29. $y = (4x + 9)^{2/5}$

30. $y = x^{-3} + x^{-5} - 6x^{-7}$

31. $y = \frac{4}{5}x^{-1} + 3x^{-3} - \frac{1}{2}x^{-4}$

32. $y = 2x^{-1/2} + 6x^{-1/3}$

33. $y = 5x^{-2/3}$

34. $y = \frac{5}{x^2} - \frac{3}{x^6}$

35. $y = \frac{4}{\sqrt{x}}$